International Series on Public Policy

The International Series on Public Policy - the official series of International Public Policy Association, which organizes the International Conference on Public Policy - identifies major contributions to the field of public policy, dealing with analytical and substantive policy and governance issues across a variety of academic disciplines.

A comparative and interdisciplinary venture, it examines questions of policy process and analysis, policymaking and implementation, policy instruments, policy change & reforms, politics and policy, encompassing a range of approaches, theoretical, methodological, and/or empirical.

Relevant across the various fields of political science, sociology, anthropology, geography, history, and economics, this cutting edge series welcomes contributions from academics from across disciplines and career stages, and constitutes a unique resource for public policy scholars and those teaching public policy worldwide.

All books in the series are subject to Palgrave's rigorous peer review process: https://www.palgrave.com/gb/demystifying-peer-review/792492.

Claudia Acciai

Policy Design for Research and Innovation

Politics, Institutions and Interest Intermediation Practices

palgrave
macmillan

Claudia Acciai
Sociology
University of Copenhagen
Copenhagen, Denmark

ISSN 2524-7301 ISSN 2524-731X (electronic)
International Series on Public Policy
ISBN 978-3-031-36627-7 ISBN 978-3-031-36628-4 (eBook)
https://doi.org/10.1007/978-3-031-36628-4

This Palgrave Macmillan imprint is published by the registered company Springer Nature Switzerland AG.
The registered company address is: Gewerbestrasse 11, 6330 Cham, Switzerland

Acknowledgment

I am profoundly thankful to Zeno, for supporting me throughout this project. I would also like to express my gratitude to my former and current colleagues for their feedback, guidance, and moral support.

Finally, I am also grateful to the Department of Sociology at Copenhagen University and to the Political and Social Science Department at the Scuola Normale Superiore, for their financial support.

Contents

1 An Introduction to Research and Innovation Policy Design 1

2 The Determinants of Policy Design Choices: A Theoretical Framework for Analysis 27

3 Choosing and Blending Instruments for R&I Policies 69

4 Research and Innovation Policy Design in France 105

5 Research and Innovation Policy Design in Italy 153

6 Comparative Research and Innovation Policy Design 205

Index 249

Abbreviations

ADEME:	*Agence de l'environnement et de la maitrise de l'énergie* (Environment and Energy Management Agency)
AERES:	*Agence d'évaluation de la recherche et de l'enseignement supérieur* (National Agency for the Evaluation of the University and Research System).
ANR:	*Agence nationale de la recherche* (National Research Agency)
ANVAR:	*Agence nationale de valorisation de la recherche* (National Agency for the Promotion of Research)
ANVUR:	*Agenzia Nazionale di Valutazione del Sistema Universitario* (National Agency for the Evaluation of the University and Research Systems)
ATOUT:	Technology Diffusion Within SME
BCRD:	*Budget* Civil de *Recherche* et Développement (Civil Budget for Research and Technological Development)
BDPME:	*Banque du Développement des Petites et Moyennes Entreprises* (Development Bank for SMEs)
BPI France:	*Banque Publique d'Investissement* (Public Investment Bank)
CDC:	*Caisse des Dépôts et Consignations* (Deposits and Consignments Fund)
CDI:	*Contrat de Développement Innovation* (Innovation Development Contract)
CDP:	*Cassa Depositi e Prestiti* (Deposits and Consignments Fund)
CEA:	*Commissariat à l'énergie atomique* (Commissariat for Atomic Energy)
CEPR:	*Comitato di esperti politiche di ricercar* (Committee of Experts for Research Policies)

CGI: *Commissariat Général à l'investissement* (Commissariat-General for Investments)
CIFRE: *Convention industrielle de formation par la recherche* (Industrial Convention for Training Through Research)
CIPE: *Comitato interministeriale per la programmazione economica* (Interministerial Committee of Economic Planning)
CIR: *Crédit Impôt Recherche* (Tax Credit for Research Expenses)
CIRST: *Comité Interministériel de la Recherche Scientifique et technologique* (Interdepartmental Committee for Scientific and Technical Research)
CIVR: *Comitato di indirizzo per la valutazione della ricercar* (Steering Committee for the Evaluation of Research)
CNR: Consiglio Nazionale della ricercar (National Research Council)
CNRS: *Centre national de la recherche scientifique* (French National Centre for Scientific Research)
CNRT: *Centres Nationaux de Recherche Technologique* (National Technology Research Centers)
CNVSU: *Comitato nazionale per la Valutazione del Sistema Universitario* (Committee for the Evaluation of the University System)
CRITT: *Centre régional pour l'innovation et le transfert de technologie* (Regional Center for Innovation and Technology Transfer)
CRT: *Centres de ressources technologiques* (Technological Resources Centers)
CSRT: *Conseil* supérieur *de la recherche* et *de* la *technologie* (Higher Council for Research and Technology)
CSRT: Conseil supérieur de la recherche et de la technologie (Higher Council for Research and technology)
D.L.: *Decreto Legge* (Decree law)
D.LGS.: *Decreto Legislativo* (Legislative Decree)
DATAR: *Délégation interministérielle à l'aménagement du territoire et à l'attractivité régionale* (Interministerial Delegation of Land Planning and Regional Attractiveness)
DGRST: *Délégation générale à la recherche scientifique et technique* (General Delegation for scientific and technical research)
DRIRE: *Direction régionale de l'industrie, de la recherche et de l'environnement* (Regional Division for Industry, Research, and Environment)
DRRT: *Délégation régionale à la recherche et à la technologie* (Regional Research and Technology Delegation)
DRT: *Diplôme de recherche technologique* (Technological Research Diploma)

EPIC: *Établissement public à caractère industriel et commercial* (Industrial and Trade Related Public Institution)
EPST: *Établissement public à caractère scientifique et technologique* (Scientific and Technological Public Institution)
EQUIPEX: *Équipements pour infrastructures de recherche / technologiques* (Equipment for Research/Technological Infrastructures)
ERAP: *Entreprise de recherche et activités pétrolières* (Enterprises for Oil Research and Activities)
FAR: *Fondo Ricerca Applicata* (Fund for Applied Research)
FAS: *Fondo per aree sottoutilizzate* (Found for Underutilized Areas)
FCPI: *Fonds commun pour l'innovation* (Mutual Fund for Innovation)
FFO: *Fondo Finanziamento Ordinario* (Ordinary Financing Fund)
FIRB: *Fondo per investimenti in ricerca di base* (Fund for Basic Research Investments)
FISR: *Fondo Speciale Ricerca Applicata* (Special Applied Research Fund)
FIT: *Fondo per l'innovazione tecnologica* (Fund for Technological Innovation)
FNS: *Fond national de la recherche* (National Fund for Research)
FRI: (Revolving Fund to Sustain Enterprises Investments)
FRT: *Fond de la recherche technologique* (Technological Research Fund)
FSI: *Fond Stratégique d'Investissement* (Strategic Investment Fund)
FSRA: *Fondo di rotazione per il sostegno agli investimenti delle aziende* (Revolving Fund to Sustain Enterprises Investments)
HCST: *Haut conseil pour la science et la technologie* (High Council for Science and Technology)
IEED: *Institute excellence sur les energies décarbonées* (Excellence Institute on Low Carbon Energy)
IHU: *Instituts hospitaliers et universitaires* (Hospital University Institutes)
IIP: *Progetti di innovazione industriale* (Industrial Innovation Projects)
INFN: *Istituto nazionale fisica nucleare* (Institute for Nuclear Physics)
IRT: *Instituts de recherche technologique* (Technological Research Institutes)
ISI: *Aide au projet d'innovation stratégique industrielle* (Support to Innovative Industrial Strategic Project)
JEI: *Jeune Entreprises Innovantes* (Young Innovative Companies)
LABEX: *Laboratoires d'Excellence* (Laboratories of Excellence)
LOLF: *loi organique relative aux lois de finances* (Organic Law on Finance Acts)
LRU: *Loi Relative aux Libertés et Responsabilités des Universités* (Law on the Freedom and Responsibilities of Universities)

MEDEF: *Mouvement des entreprises de France* (Movement of Companies from France)
MESR: *Ministère de l'Enseignement supérieur, de la Recherche* (Ministry of Higher Education, Research)
MINEFI: *Ministère de l'Économie et des Finance* (Ministry of the Economy and Finance)
MIRES: *Mission interministérielle Recherche et Enseignement supérieur* (Interministerial Mission for Research and Higher Education)
MISE: *Ministero dello sviluppo economico* (Minister of Economic Development)
MIUR: *Ministero dell'istruzione università e ricerca* (Minister of Higher Education and Research)
OMC: Open Method of Coordination
PFT: *Plateformes technologiques* (Technology Platforms)
PIA: *Investissement pour l'avenir* (Investment for the Future Plan)
PPA: *Prêt Participatif d'Amorçage* (Participatory Priming Loan)
PRES: *Pôles de recherche et d'enseignement supérieur* (Research and Higher Education Clusters)
RDT: *Réseau de développement technologique* (Technological Development Network)
RRIT: *Reseaux de Recherche et d'Innovation Technologique* (Technological Research and Innovation Network)
SAAT: *Société d'accélération du transfert technologique* (Technological Transfer Acceleration companies)
SAIC: *Services d' activies industrielles et commerciales* (Commercial and Industrial Business Services)
SGR: *Società gestione risparmi* (Asset Management Company)
SLR: *Sauvons la recherche* (Let's Save the Research)
SNRI: *Stratégie Nationale de Recherche et d'Innovation* (National Strategy for Research and Innovation)
SUIP: *Société Unipersonnelle d'Investissement Providentiel* (Unipersonal Risk Investment Company)

List of Figures

Fig. 2.1	Instrument families	41
Fig. 2.2	Instrument shapes (by authoritative dimension)	44
Fig. 2.3	Instrument delivery component	45
Fig. 2.4	Exploratory framework of analysis	47
Fig. 3.1	GBARD statistics France	73
Fig. 3.2	GERD statistics France	74
Fig. 3.3	GBARD statistics Italy	75
Fig. 3.4	GERD statistics Italy	76
Fig. 3.5	Typology of institutional layouts	78
Fig. 3.6	The variety of intermediation strategies	80
Fig. 3.7	The methodological strategy	82
Fig. 4.1	The chronology of R&I policies (France)	106
Fig. 4.2	Recipients of the Tax Credit for Research by enterprise size	128
Fig. 4.3	Share of Tax Credit for Research benefits by enterprise size	128
Fig. 5.1	The chronology of R&I policies (Italy)	154
Fig. 6.1	The chronology of the two cases	208
Fig. 6.2	Distribution of instrument families by cabinet political orientation	212
Fig. 6.3	Distribution of instrument shape coerciveness by cabinet political orientation	213
Fig. 6.4	Distribution of delivery component characteristics by cabinet political orientation	214
Fig. 6.5	Aggregate public expenses on R&I mission by minister	223
Fig. 6.6	Choices of instrument delivery component by country	235

List of Tables

Table 2.1	Policy design literature: a focus on instruments	30
Table 2.2	Policy Instrument classifications	37
Table 3.1	List of interviewees	89
Table 3.2	Codebook and description of the contents	90
Table 4.1	Instrument mix—law on innovation and research	140
Table 4.2	Instrument mix—innovation plan	142
Table 4.3	Instrument mix—investment for the future plan	143
Table 4.4	Instrument mix—research act	145
Table 4.5	Business-oriented policy mix I (1980s–1990s)	146
Table 4.6	Business-oriented policy mix II (2000s)	149
Table 4.7	Cluster I—instrument mix	150
Table 4.8	Cluster II—instrument mix	151
Table 5.1	Instrument mix—*Bassanini reform*	187
Table 5.2	Instrument mix—plan for digital innovation in enterprises	189
Table 5.3	Instrument mix—*Industria 2015*	191
Table 5.4	Reform in public research sector I—instrument mix	193
Table 5.5	Reform in public research sector II—instrument mix	193
Table 5.6	Business-oriented I—instrument mix	194
Table 5.7	Business-oriented II—instrument mix	194
Table 5.8	Business-oriented III—instrument mix	196
Table 5.9	Business-oriented IV—instrument mix	197
Table 5.10	Business-oriented V—instrument mix	198
Table 5.11	Business-oriented VI—instrument mix	199
Table 6.1	Ministerial coordination initiatives: France and Italy	219
Table 6.2	The ministerial contribution to MIRES (2012)	221

CHAPTER 1

An Introduction to Research and Innovation Policy Design

1.1 Introduction

National economies are moving toward new technological paradigms, where information and communications technology (ICT), green tech, and other computational technologies are increasingly acquiring a prominent role in national policy agenda. Similarly, the diffusion of innovative technologies is transforming not only the way our societies function but also the way policymakers deliver policies and interact with stakeholders (Pencheva et al. 2018). Over the last few decades, despite contrasting interpretations on the drivers of economic development, a widespread agreement that innovation and knowledge development represent important components of economic growth has emerged and consolidated. Their contribution to growth can be found in the enrichment resulting from technological progress embodied in physical capital (investments in more advanced machinery or computers), in the results coming from investments in intangible capitals (software, design, data, firm-specific skills), and in supplements linked to the increased efficiency in the use of labor and capital (OECD 2015b, 17).[1] Hence, when we analyze the development of a country's national system of Research & Innovation

[1] Some of the policy fields in which OECD (2015b) registered an accelerated technological progress are the following: climate change mitigation, aging, health, food security, information, and communication management.

C. Acciai, *Policy Design for Research and Innovation*, International Series on Public Policy,
https://doi.org/10.1007/978-3-031-36628-4_1

(henceforth, R&I), we are reading this country's economic, social, and political history through the prism of the particular conditions surrounding the use of knowledge and technology in economic production, together with the choices made by the ruling class (or dominant elites), regarding the production and application of scientific and technological knowledge (Chesnais 1993, 194). From this perspective, we can interpret R&I policies as a public investment in long-term intangible assets able to generate profits in the future. This is different from regular public investment strategies, such as capital expenditures, because of its longer time horizon and higher risk (Bhattacharya et al. 2017). It entails a vision of national R&I strategies linked with the production of knowledge and its empirical applications to solve different problems that are potentially related with all the tasks of government. One only needs to think about the different research centers, or public agencies, attached to various functional ministers (e.g., agriculture, health, environment) to get a hint about the potential relevance of R&I investments for the design and implementation of public action (OECD 2014). Therefore, the meaning of innovation has also evolved, moving from a means to achieve a broad range of policy goals, toward a goal for a broad range of policy sectors (Flanagan et al. 2011).

State promotion for research and technological development has a long tradition which dates back to the pre-World War II period, when public research investments were mainly oriented toward agriculture (plant breeding, plant protection), health (vaccines), mining (geology, engineering), or navigation, transport, and communication (shipbuilding, aeronautics, and telecommunications) (Nauwelaers and Wintjes 2008). And, similarly, the COVID-19 health crisis has highlighted how investments in scientific research and the capacity to develop new technologies represent essential assets for contemporary societies. In response to all these challenges, the scope of R&I policies has broadened beyond its traditional missions (Meissner and Kergroach 2019); and the governance of the sector has also evolved. Private actors, through hybrid combination with public research actors, have progressively gained greater prominence within the field of scientific and technological activities used in the economy (Arnold and Boekholt 2003). Meanwhile, policymakers are increasingly attempting to design more holistic strategies aimed at translating knowledge and technological advancements into policy actions. This scenario designs a governance arrangement where actors are interdependent while competing for their own stakes and interests. Namely, policymakers need knowledge advancements for responding to pressing societal issues while knowledge and technology producers require some form of public support to produce them.

This research aims at investigating how these apparently contrasting powers (policymakers versus policy takers) and interest struggles influence the design of R&I policy interventions in a governance arrangement characterized by a high degree of interdependency between actors. This study adopts a novel perspective on the study of policy design by analyzing this process from the perspective of both the actors "making" and the actors "receiving" policy design choices. This perspective allows us to investigate policy design from a two-way perspective, where both policymakers and policy takers have the power to influence the likelihood of certain instrument features to be selected. First, focusing on national decision-makers, this book explores how national political priorities have been translated into R&I policy design choices. Then, zooming in on those actors involved in the policy design of national R&I strategies, it investigates the way different institutional structures, and their related coordination challenges, influence policymaking activities. Finally, examining policy recipients and the way these interact with public institutions, the study investigates to what extent actors' policy responsiveness shapes policy design choices.

Therefore, the innovative contribution of this research is twofold. First, the book adopts an actor perspective in the study of policy design with a spin in the direction of including behavioral insights into the study of policymaking. This perspective is especially relevant for R&I policies, where policy design choices encompass interdependent issues related to knowledge production and exploitation practices, that do not necessarily map easily onto the typical remits of formal policymaking institutions (Pelkonen et al. 2008). By focusing on actors, their interactions, and stakes, the book unravels the complexities characterizing policymaking in a complex governance environment, like R&I policy. Secondly, the book introduces a new treatment of policy instruments and then applies this heuristic for investigating the operational dimensions of R&I national governance arrangements. This perspective disentangles the different pathways shaping R&I policy design choices in France and Italy.

Following Lasswell's definition of politics as "the process of who gets what, when, and how," the research deciphers the extent to which actors' choices and interactions matter for policy design and how alternative political, institutional, and interest intermediation configurations can influence this process. Hence, the innovative contribution of this analytical approach is precisely to investigate policy design from the perspective of both the actors "making" and the actors "receiving" those decisions.

Ultimately, this also contributes to a deeper understanding of how governments design more ambitious R&I policy mixes aimed at solving the cross-sectorial social challenges our contemporary societies are facing.

The book develops these arguments across a comparative case-study analysis of two Western European countries' innovation systems: France and Italy. From the perspective of R&I policies, these countries are two apparently similar and binary systems; nevertheless, within this similarity, there are also very different approaches characterizing policymaking. Indeed, when looking at the aggregated characteristics of their instrument mix (e.g., how social control is exercised and the relationship between policymakers and target population) France and Italy display a greater variety. Such differences reflect alternative national strategies to interact with target populations, as well as the different extents of activism and the organizational capacity of national R&I performers. Therefore, despite both countries sharing a stable center-of-government R&I structure, and a similarly sophisticated instrumentation, they still display some heterogeneity in their R&I instrumentation. Hence, in understanding and explaining these differences, the book introduces its core contribution to the policy design literature, suggesting how the type and the way instruments are selected and designed (rather than only their quantity) matters for policy changes.

To conclude, Chap. 1 is organized as follows. The coming section introduces the policy sector under investigation and delineates its conceptual and operational boundaries according to the current scholarship. Section 1.2 contextualizes policymaking in governance arrangements characterized by policy area distinguished by the sectorial dispersion of public responsibilities. The following three subsections introduce the core pillars adopted in the analysis of R&I policy design throughout the book, namely partisan politics, institutional layouts, and policy takers' responsiveness. Finally, Sect. 1.3 presents an overview of the book's contents.

1.2 Policymaking in Complex Governance Environments: The Case of R&I Policies

The field of R&I policies is a highly multidisciplinary area and "much of what is called innovation policy today may previously have gone under other labels such as industrial policy, science policy, research policy, or technological policy" (Edler and Fagerberg 2017, 5). Therefore, it is necessary to clarify how this research conceptualizes innovation and the related system of public activities supporting its development. The Oslo Manual[2] identifies four types of innovations, namely: organizational innovation, market innovation, process innovation, and product innovation.[3]

Hence, if inventions can be defined as the first occurrence of an idea (and its first application), innovation represents the systematic development and exploitation of this idea into practice (Fagerberg et al. 2009). However, the set of activities which stand at the upstream of innovation processes are also relevant; as defined by the Frascati Manual,[4] these are classified as follows: basic research, applied research, and experimental development.[5] Indeed, knowledge production and development are often necessary, although not sufficient in themselves, factors for producing innovation.

Countries differ in their economic and industrial specializations; similarly, they also tend to display different needs with respect to their demand of skills, knowledge, and finance structures. Similarly, the interaction between policymakers' goals, knowledge infrastructures, and innovation performers are consistently shaped by these needs (Edler and Fagerberg 2017). Thus, the governance of R&I defines a system of interactions between the actors involved in knowledge production and knowledge application sectors; hence, between the state, firms, university, and public research organizations (Amable and Petit 2001; Considine et al. 2009; Edler et al. 2016; Edquist 2001a). Consequently, these two systems

[2] It is the OECD document adopted as the international reference guide for collecting and using data on innovation.

[3] Organizational innovation refers to the implementation of a new organizational model or business practice; market innovation concerns the introduction of a new marketing method involving changes in product design and product placement; process innovation is the implementation of a new method of production; and finally, product innovation refers to the new improvements in technical specification, components, and material for both goods and services (OECD/Eurostat 2018, 45).

[4] It is the OECD document adopted as the international reference guide for collecting statistics about Research & Innovation.

[5] Basic research is experimental or theoretical work undertaken primarily to acquire new knowledge of the underlining foundations of phenomena and observable facts, without any particular application or use in view. Applied research is an original investigation undertaken in order to acquire new knowledge; it is, however, directed primarily toward a specific practical aim or objective. Experimental development is systematic at work, drawing on knowledge gained from research and practical experience and producing additional knowledge, which is aimed at producing new products or processes or at improving existing products or processes (Frascati Manual 2015a, pp. 50–51).

should be considered jointly since they mutually shape each other's needs and features. This reasoning is also consistent with the approaches found in the literature, which focuses on adopting a wider approach for investigating the origins and dynamics of innovation (Smits et al. 2010). This study delineates the operational boundaries of *Research and Innovation Policies* as:

> the system of public activities supporting research and knowledge production, and their coordination with public interventions focused on the development and exploitation of this knowledge for various goals (e.g., economic growth, national competitiveness, employment, grand societal challenges).

Research policies focus on the generation of new ideas and innovation policies on the activities focused on the exploitation of these ideas into practice (Edler and Fagerberg 2017). These policies embrace instances of both requests to support public research (Cantner and Pyka 2001) and the demand, development, and application of this knowledge into practice (Borrás and Edquist 2019).

Consequently, public interventions in R&I often address broad challenges that, given the impossibility of their being tackled as a whole, demand to be translated into smaller scale issues (Edquist and Zabala-Iturriagagoitia 2012). Multiple ministers and stakeholders can be potentially involved in policymaking processes precisely because R&I issues stand at the crossroads of many established sectors of policy interventions, which are usually institutionalized according to different ministries, public agencies, or specific administrative departments. Therefore, R&I policies design multi-level, multi-actor, and multi-issue governance systems (Chou et al. 2017) where each intervention needs to be joined up across a broad set of delivery areas including tax, science, education, immigration, enterprises, foreign and direct investments, and health policies. For these reasons, R&I policies represent an emblematic "case of" (Ragin and Becker 2009) complex governance arrangement, where policymakers are increasingly being asked to solve problems that cannot be easily categorized into one policy sector. Similarly, this phenomenon is also influencing other "classic policy areas" challenging national policy agendas with more complex and cross-sectorial issues, like climate policies (Van Asselt et al. 2015), migration policies (Scholten et al. 2017), and health policies (Trein 2017). Their common denominator is that all these policies aim at addressing grand societal challanges like, for example, climate change, food security, and other economic, societal, and technological threats (WEF 2020) which cannot solely rely on the design capacity, and resources, of one unique policy area. In these policymaking contexts the burden of design choices is increasingly shared across different policymakers. Consequently, in order to gather the necessary resources and design holistic R&I policies, new institutional configurations have to be found among different policy responsibilities (Borrás and Edquist 2019). However, enhancing coordination, and collaboration, across different policy sectors is not so easy. This is because specific policy responsibilities (and sectors) are characterized by relatively stable groups of actors and institutions, associated interests, representation practices, and perceptions of problems which can further hinder opportunities for collaborations (Candel and Biesbroek 2016).

At this point an important question is whether existing policy design theories are well equipped to interpret this complexity. The current scholarship describes how different governance structures define alternative systems of integration across "classical" sectors of policy responsibilities. This can happen through the implementation of specific organizational layouts which can reflect specific national framings of R&I issues, what is defined as boundary-spanning policy regimes (Jochim and May 2010) or horizontally connected policy sectors (Trein 2017). Therefore, policy integration represents a political goal that must be obtained (Tosun and Lang 2017, 560) and this can either rely on similar framings of policy problems (Nilsson and Nilsson 2005) or it can be the result of conscious organizational designs based on strategic considerations (Christensen et al. 2014). In his study on health care and public health policies, Trein (2017) analyzes an emblematic case of integration between policy sectors. Here, the author shows how the governance capacity to coordinate the actions of different policy subsectors is dependent upon the institutional design adopted for integration (e.g., the ministerial structure) and the collaborative attitudes of stakeholders involved, so-called actors' responsiveness. Other contributions (Candel and Biesbroek 2016) suggest how policy integration challenges become particularly severe when complex societal issues are confronted with governance systems characterized by relatively stable actor configurations, associated interests, and problem perceptions.

Policy design is fundamentally an intellectual process of constituting a reality and then attempting to work with it (Linder and Peters 1998, 45). Therefore, design choices are usually linked to the general policy preferences dominating a specific sector (Cashore and Howlett 2007), whose dynamics are also influenced by the way intersectionality and integration take place. However, as discussed in the policy instrument literature in democratic societies, no single actor possesses sufficient resources to unliterally influence policy change, and governments are increasingly dependent upon the action and acquiescence of others they do not directly control (Bressers and O'Toole 2005; Capano and Lippi 2017; Majone 1976; Ringeling 2005; Woodside 1986). For all these reasons, policy instruments ultimately represent the building blocks of alternative governance agreements between policy takers and policymakers (Capano et al. 2019).

The present study adopts this reasoning and, going backward, investigates how policy design choices shaping specific governance arrangements have been influenced by the relationship between policymakers and policy takers. Through a reconstruction of the R&I national policy design process, the rationalities of the actors involved, their interests and strategies for action, the analysis provides a deeper understanding of how the interactions between differently motivated actors shape the characteristics of national R&I strategies. Therefore, we aim at moving forward into this field of research and to investigate the influence that different configurations of cross-sectorial governance arrangements exert on R&I policy design choices. The value added by the present study is in the adoption of an interpretative framework of analysis considering different rationalities for state intervention, the institutional capacity of policymakers to integrate ministers with complementary competences, as well as the role target populations play in easing, or hampering, these practices.

1.2.1 Different Rationalities for State Interventions: The Politics of Policy Design

Policymakers' attention toward knowledge-intensive activities is related to the positive spillovers these can trigger in different policy areas. R&I does often represent a means to achieve long-term political goals like economic growth and competitiveness, employment, environmental protections, and, more broadly, strategies to tackle grand societal challenges. Hence, a country's political context is said to (in)directly influence the availability of resources for investments in R&D and to differently legitimize spaces in which firms can gain access to resources (Van de Ven and Garud 1989). This is because technological progress and, more broadly, knowledge advancements represent key factors necessary to sustain economic and social development.

The history of R&I policymaking can be divided into four different phases, describing alternative logics of priority setting. The first refers to the aftermath of World War II (1940s–1950s). Here policymakers' goals, in what nowadays is called R&I, were focused on key military technologies, with the aim of supporting the production of public goods in areas where there was no private incentive to do so. Their primary aim was restricted to developing particular technological capabilities (Foray et al. 2012), with a purely top-down, market-fixing rationality (Kattel and Mazzucato 2018; Laranja et al. 2008). In this context, state intervention was understood as being a necessary condition to overcome market failures embedded within the public good characteristics of knowledge. The following period was distinguished by the application of R&I goals for civilian industrial technologies, and this was contextual to the post-war industrial and economic development era. The dominant framing was mainly driven by a first-mover advantage logic, and it focused on the development of private industry with the ultimate goal of disseminating results to the wider public (Gassler et al. 2008). The innovation process was mainly interpreted as a linear mechanism starting from basic research, through applied research and development, up until the market introduction of new products and technologies (OECD 2005). Public actors played a prominent role in leading innovation processes often from the outset, through public research laboratories, either military or civilian, until the output, by supporting (or, in some cases, owning) national industries. The third phase developed across the 1980s and it was closely associated with the emergence of the National System of Innovation (NIS) approach (Freeman 1987; Gassler et al. 2008; Lundvall 1992; Nelson 1993). Here, the innovation process was understood as being the result of the interplay between systemic elements with institutional framework conditions; while the production of innovation was made up of constant interactions and learning between the members of this network (Edquist 2001b). Public intervention was perceived as a necessary, although not sufficient, factor in supporting coordination to ease the cooperation between actors and organizations involved in a specific network. The rationality for state intervention was guided by the perceptions of different systemic failures related to the network-based nature of the sector, such as infrastructural failures, capability failure, and institutional and network failures (Chaminade and Edquist 2006; Cimoli et al. 2009; Edquist 2001a, 2014). Consequently, public intervention was meant to be (sometimes) required, but it couldn't nevertheless completely displace the role of private actors. The last period, which began around the early 2000s, saw the role of R&I policies as increasingly oriented toward responding to major societal and economic challenges.[6] After the adoption of the Lisbon strategy (2000–2010), this narrative has further institutionalized the framing of R&I as an essential component for national economic growth (Ulnicane 2016). With the grand societal challenges policy approach, R&I policies have started to be directed toward pressing issues for societal developments (e.g., environmental and health problems, job creation, and security and defense) (Borrás 2009; Gassler et al. 2008). Such an approach has increased policymaking attention toward how R&I can be instrumental for new strategies pursuing collaboration across different policy sectors in the effort to tackle contemporary policy problems (Head and Alford 2013; Kattel and Mazzucato 2018; B Guy Peters 2017). This historical overview shows how alternative problem-solution framings can actually cluster into long-term R&I rationalities (Edler et al. 2016; Fagerberg 2017), acting as a lens that filters information and directs decision-makers' attention toward different issues (Wilson 2000). However, alternative logics for priority setting in R&I do not substitute their predecessors, but they rather develop new features, which add up to the existing structures of priorities (Gassler et al. 2008, 220). Therefore, the extent to which policymakers adopt one or the other rationality is something that can only be assessed empirically (Béland 2009; Palier 2007).

The scholarship has demonstrated how the deployment of R&I instruments is able to alter the space of politics (Edquist and Borrás 2013). The extent to which the political context can differently influence the availability of financial resources for investments in R&D (Bergek et al. 2015), while legitimizing different spaces where policy recipients can gain access to resources (Van de Ven and Garud 1989) is of primary concern here. Also discussed in the literature is the influence of left- versus right-wing party cleavages on the shape of university tuition fees (Kauder and Potrafke 2013), as well as on the pace of technological innovation (Wang et al.

[6] This concept has been used for the first time in 2007, in the context of the Green Paper on the European Research Area.

2019). Other contributions focus on the impact that political uncertainty plays on the attitudes toward innovation of different governments (Ma 2017) and on the economic incentives to innovate (Bhattacharya et al. 2017). However, despite a growing scholarly interest in R&I instrument mix, and more broadly in policy design, there is little evidence of how instrument mixes are blended (Ghazinoory et al. 2019; Meissner and Kergroach 2019; Neicu et al. 2016). The possibility for new material conditions to trigger policy change is dependent upon the interpretative ability of existent policy frames to provide convincing solutions to "new problems" (Sanz-Mendez and Borrás 2001, 5). The formulation of R&I policies represents a political strategy by means of which decision-makers are able to identify problems and the way they want innovation to contribute to their solutions (Edler and Fagerberg 2017). Hence instruments, and the way they are implemented, are the result of a political process, with the tool ultimately being a manifestation of a dominant worldview and influence (Edler et al. 2016, p. 8). Both individual values and long-lasting assumptions about how the world should be are key components in shaping the politics of problem definition. Thus, translating instrument choices into R&I intensity targets is the result of a specific political framing (Edquist and Borrás 2013), which tends to be influenced by both contingent perceptions and long-lasting assumptions about how the world should be. All of this suggests how two counteracting forces are at play when choosing and blending instruments, namely: the different partisan orientations of the cabinets in office and the legacy of consolidated types of R&I national policies.

When selecting new instruments, policymakers are caught in the grips between their current preferences on how to shape new instrument mixes and the effective room for maneuver they have given the distribution of power set in place by previous choices. The choices on how to translate their political priorities into research and innovation-intensity targets are the result of a specific political framing, which tends to be influenced by both individual values and perceptions, as well as by long-lasting assumptions about how the world should be. Therefore, in order to understand the way instruments are blended into mixes, and the drivers behind their selection, it is important to investigate how partisan politics, in its combined effect with path-dependency forces, influences policymakers' choices over an alternative instrument mix. Policy instruments are chosen with a purpose (Edler et al. 2016); so, the way these are designed and implemented represents a manifestation of different political views. Hence, investigating the interaction between politics and the consolidated national R&I policy styles can explain the mechanisms shaping R&I policy goals' definition. More specifically, we seek to answer the question: How do the political orientations of different cabinets influence the formulation of national R&I policy strategies? Therefore, we aim at investigating the way different political priorities have been translated into R&I instrument mixes.

1.2.2 The Institutional Shapes of Integrated Policy Sectors

R&I policies do not map onto the typical areas of activity of formal policymaking institutions (Pelkonen et al. 2008). This requires the institutional, and organizational, capacity necessary to cope with issues at the intersection of "classic policy sectors." A practice that often comes together with an increased blurring of competences between executive national departments. For this reason, R&I policies represent an emblematic "case of" (Ragin and Becker 2009) complex governance arrangements, in which policymakers are increasingly being asked to solve problems that cannot be easily categorized into one policy sector. This research adopts a strictly institutional approach (Edler et al. 2016) to investigate national R&I policymaking processes, hence focusing only on the study of policymaking organizations (ministers or public agencies), with direct competences in the R&I sector.[7] From an institutional organizational perspective, R&I policies, such as climate policies (Van Asselt et al. 2015), migration policies (Scholten et al. 2017), and health care policies (Trein 2017), stand at the intersection of "classic policy sectors," mirroring different functional executive policymaking organizations. Consequently, the organization of R&I policymaking competences, and the institutional system that supports them, can vary according to the political context or national public administration traditions (Edler et al. 2016; Edler and Fagerberg 2017). This is to say that the way in which R&I policy responsibilities have been organized reflects different national framings of R&I issues and of their related policy strategies (Tosun 2018). Since ministerial organization is not strictly

[7] We did not focus on the measures in support of R&I designed and implemented by functional ministries and agencies (e.g., energy, health, or transport), serving the purpose of supporting innovation as a means to achieve their ultimate policy goals (Edler et al. 2016). That is because in this research we are interested in the formal policy processes aiming to get the various institutional and managerial systems involved in R&I issues to work together (OECD 2005: 9).

determined by law, institutional organization structures (for the most part) tell us how national governments have differently framed their policy problems and the political priorities to be tackled (Peters 1998). Ministerial organizations retain a central role in the formulation of policies within their areas of competences given their expertise, resources, and institutional power in the policymaking process (Laver and Shepsle 1996). The literature recognizes the influence of different institutional layouts on the ability of research and industrial actors to produce knowledge and innovation, as well as of policymakers to invest and regulate (Braun 2008a; Chung 2013; Smits et al. 2010). For instance, in describing the Danish case, Koch (2008) shows how changes in ministerial organizations, and in their internal dynamics of coordination, are related with shifts in the instruments characterizing the national R&I policy mix. Similarly, Edler and Kuhlmann (2008) in their analysis of the German innovation system, explore the relevance of ministerial coordination practices for designing a balanced (in terms of a coordinated blend of different policy measures) R&I policy mix, within and across knowledge sectors. Finally, Griessen and Braun (2008) also describe how despite Switzerland being grouped among the most successful countries in terms of knowledge production and technological innovation, the country still faces consistent cross-sectoral policy coordination challenges within its institutional structures. This suggests how different institutional layouts come with different trade-offs, according to how the actors managed to integrate different policymaking structures and interact with policy recipients. The organization of R&I institutional layouts requires striking a balance between coordination (e.g., designing an all-encompassing policy strategy) and specialization (e.g., the calibration of instruments to specific contexts and recipients). However, due to the different systems of competence division and related mechanisms of habituation and socialization within organizational structures (Binder et al. 2009; Merton 1938), there is no best organizational strategy (Pelkonen et al. 2008). Different layouts influence the way authority is exercised, meaning how ministers and public agencies organize their actions in the governing business. Departmental identities affect selective perceptions influencing actors' attention toward different phenomena while shaping their understandings of causal effectiveness (Scharpf 1997, 40). However, the combination between R&I issues' complexity (Verhoest and Bouckaert 2005), the traditional siloed approach characterizing public administrations (Christensen et al. 2014; Pencheva et al. 2018), as well as the classic departmentalized structure of formal policymaking institutions (Pelkonen et al. 2008) call for some degree of coordination.

The literature recognizes the influence that inter-institutional networks and different coordination practices exert on the ability of research and industrial actors to produce knowledge and innovation, as well as of policymakers to invest and regulate (Braun 2008b; Chung 2013; Edler and Kuhlmann 2008; Griessen and Braun 2008; Koch 2008). Our contribution builds on this evidence and aims to move the focus of analysis forward, to investigate how the different morphologies of ministerial coordination practices influence the ability of policymakers to design instrument mixes capable of considering multiple traditional subsectors of public action. We still know little about how this influences the way R&I instruments are designed, and especially the blending of rationalities, behavioral goals, and implementation styles administered to policy recipients. Hence, this study aims to investigate how the morphologies of ministerial organizational layouts influence the ability of policymakers to integrate different policy responsibilities.

The analysis focuses on exploring the extent to which the internal policymaking ability in R&I was challenged by the underlining necessity to integrate different policy responsibilities.

1.2.3 A "Policy Taker" Perspective for Understanding Policy Design Choices

In policymaking activities requiring technical competencies and skills decision-makers cannot provide by themselves, like the case of R&I policies, public actors are likely to rely on the collaboration of external actors. During this delegation of responsibilities, the agent should perform certain agreed-upon tasks, with some consequential benefits accruing to the principal (policymakers) as well (Guston 1996, 230). This process embeds a trade-off between freedom and public (policy) obligations related to the role of science and technology for societies. In the literature the necessity to preserve the autonomy of science as a premise for its development and circulation among researchers, governments, industries, or elsewhere is well consolidated (Bush 1945; Merton 1938). However, as increasingly suggested by other contributions from the literature (Mazzucato 2013), there are several experiences where the role played by state actors has been pivotal for the development of new societal, economic, and technological innovations. In some national instances, the creation of intermediary institutions between policymakers and R&I performers has represented a strategy to stabilize the interest of the actors involved and to temper the politics-science conflict. For instance, in the Danish case featuring the creation of different research councils with tasks related to the coordination of public research, the production of policy advice, and, partly, the management of national funding allocation represented a strategy to balance university autonomy (Koch 2008). However, as demonstrated by the comparative case study of Braun (1993), these strategies do also represent a double-edged sword. On the one hand, the creation of intermediary agencies with competences in R&I policy has relaxed administrative burdens by integrating access to information and cooperation with recipients. However, on the other hand, this has granted considerable power to R&I performers.

Policymakers have a stake in influencing the way R&I goals can become instrumental for their own political (and economic) goals. Consequently, it becomes difficult to identify those actors who can achieve the necessary legitimacy, and power, to determine national R&I investment decisions and whether these choices should be driven by market needs versus scientific considerations.

One of the core themes in the design of science policies concerns how public action can best support research and technological development with public funds. While, conversely, this is also related to the ways policymakers can support scientific and technological development without undermining its vigor, initiative, and independence (Ben-David 1991). Consequently, some extent of preference alignment, or intermediation, between the principal and the agent is needed. This is what Trein (2017), in analyzing the same phenomena in health policy, defines as "actors' responsiveness."

When different actors are cooperating in the production of a public good, the major problem is how to shape their opportunity structures in order to limit their tendency to "free-ride" (Braun 1993, 137–38). Consequently, policy preferences of target populations (policy recipients) also become important, to the extent that the lack of stakeholder support, or legitimacy, for the behavioral change sought by instruments can hamper cooperation and undermine successful implementation (Curley et al. 2020; Dermont et al. 2017; Gross 2007; Ingold et al. 2018; Kammermann and Ingold 2019; Varone and Aebischer 2001). All of that suggests how it has increasingly become necessary to consider that governments are no longer simply and unilaterally authoritative, but they are instead dependent upon the action, acquiescence, or support of others, which they do not directly control (Bressers and O'Toole 1998; Goetz 2008; Salamon 2002). This perspective requires an element of "supply push" in analyzing policy design, which allows us to understand how policy takers can influence policy design. We know that instruments may influence, in a distinct fashion, the way target groups think about their own behaviors and choices (Schneider and Ingram 1993) and that the way policy takers use these instruments influences the outcomes of a policy (Curley et al. 2020). Policy takers do benefit from public funding (in different forms) for their activities, but policymakers also benefit from the results of these activities. Consequently, current challenges in understanding R&I policy design processes require disentangling the relationship existing between the actors involved in the broader governance of the sector, on the part of policymakers and policy takers, respectively. Our understanding of how R&I policy recipients influence the policy design process is still limited. Policy design is increasingly exposed to horizontalization pressures (OECD 2005), since both top-down and bottom-up forces are present, and different actors actively negotiate over the design of national R&I policy strategies (Gassler et al. 2008). This is related to the fact that in R&I activities, investments from both public and private actors are required (Kattel and Mazzucato 2018). Therefore, it becomes increasingly necessary to understand the role of policy takers in R&I policy design and to what extent different strategies for interest intermediation influence R&I instrument mix choices. This means focusing on exploring the way actor responsiveness shapes the policy design process.

1.3 Aims and Scope of the Book

This book investigates the determinants of policy design choices in R&I. It explores how governments design national R&I policy strategies, both internally (institutional coordination of sectorial responsibilities) and externally (in their relationship with policy takers), by analyzing the influence of different governance configurations on policy design choices. Focusing on national decision-makers, it investigates the role of different partisan politics orientations in shaping their choices for alternative instrument features. Then, narrowing the focus on those actors involved in the policymaking of national R&I strategies, it explores the different ways institutional structures and their related coordination challenges influence policymaking activities. Finally, the book focuses on the interaction between policymakers and policy takers, investigating how their configurations influence policy design choices.

The core idea behind this study is that not only the type of instrument matters but also its capacity to induce a specific behavior while addressing the appropriate target. Similarly, the mix of instruments created by specific policy design choices makes a difference regardless of the quantity of the instrument shapes included. Despite the fact that instruments may look similar in the way they deal with a problem, there will always be substantial differences in the concrete details of how they are chosen and designed in the context in which they are applied. Policy instruments are political devices because they display a high degree of interpretative flexibility, due to the stratification of new instruments and rationales (Flanagan et al. 2011). Hence, analyzing policy instruments from the perspective of the behavioral changes required for the target population can better support our understanding of the policy design process as well as their stratified influence on the dynamics of the R&I sector. This study aims at understanding the way different policy tools can affect the reality where actors interact; our analysis is focused on a deeper understanding of how social control is exercised (instrument shape) and of the types of governing arrangements coming together with the selection of different instruments (delivery structure). By introducing a theoretical and analytical differentiation between instrument families, shapes, and delivery components, we aim at overcoming the (restricted) focus on policy instruments as a means by which governments use their legitimate power to shape public action (instrument families). Exercising power means to obtain collaboration, that is inducing behaviors in line with our expectations (Stoppino 2001). To understand how authority is effectively exercised, we need to make sense of the different relationships policy instruments (or combinations of thereof) establish between policymakers and target populations. By introducing the analytical category of instrument shape, we are indeed able to grasp the substantive characteristics of the inducement effectively administered to the target population through the selection of a specific tool. Meanwhile, the delivery component provides information on those actors who have the titularity, and the power, to steer instrument action along the management process (Bouwma et al. 2017). Therefore, by adopting the new treatment of R&I policy instrument typology proposed in our research, we aim at providing a greater understanding of the behavioral and political characteristics embedded in instrument action; while, at the same time, showing how both the choice and the particular way instruments are crafted reflect different types of interactions between decision-makers and policy recipients.

The research presented here addresses these questions with two innovative contributions. First, it analyzes the determinants of R&I policy design choices from the perspective of decision-makers by illustrating how policymakers interact across institutions and with R&I stakeholders. This represents a novel contribution to the literature because it combines the supply-side with the demand-side of policy design. Indeed, this book analyzes policy instrument choices from the perspectives of both policymakers and policy takers. Secondly, to interpret this biunivocal relationship, the author theorizes a new R&I policy instrument treatment relying on instrument families, shapes, and delivery components. This classification differentiates instruments according to the behavioral changes required for policy takers and it illustrates the different ways in which public authority is exercised. The book applies this heuristic for the analysis of R&I policy design, and it demonstrates how the capacity of different instruments, and mixes, to induce behavioral changes influences national policy design strategies. The author tests these arguments through a diachronic investigation of the evolution of R&I policy instrument mixes across two European countries characterized by diverging national policy design strategies: Italy and France. Both cases share a stable center-of-government R&I policy mix structure, especially when focusing on the way these governments use their power to steer target populations toward the intended behavior. However, when looking at the stratified characteristics of their instrument mixes, and especially at the relationships these created between policymakers and policy takers, the comparative analysis demonstrates greater variety. These differences reflect alternative approaches the two countries have undertaken to interact with target populations, as well as the various geographies of political entrepreneurship and the organizational capacity of national R&I performers. The author builds on this comparative evidence to demonstrate how the type and the way instruments are selected and designed (rather than only their quantity) matters for policy changes. By focusing on the interaction between policymakers and policy takers in the French and the Italian R&I governance systems, it investigates the way actors, in their interactions with a political and institutional context, shape the design of national policy strategies. The remainder of the book is structured as follows. Chapter 2 introduces the readers to the conceptual toolkits adopted throughout the analysis. It first touches upon the extant literature on policy design, reviewing the scholarship on policy formulation, combined with a specific angle on policy instruments. Then it moves on by presenting the proposed policy instrument typology together with the analytical framework adopted for the analysis. This framework is also discussed in Chap. 3, where the theoretical concepts are applied to the two cases under investigation. Here the author also presents the methodology and data sources adopted in the analysis, while also providing a more general introduction to the case studies of each country. This chapter provides a review of the existent literature on the study of policy formulation in a sector characterized by policy issues which do not necessarily map easily onto the typical remits of formal policymaking institutions. Chapters 4 and 5 dive into the analysis of the two case studies, adopting a parallel structure, combining a historical perspective with evidence and references to official policy documents (for more details please refer to the Appendix for each of the two chapters). In the former, the book describes the diachronic evolution of French R&I policymaking from the angle of the different policy instruments adopted. The latter draws on the empirical analysis of the Italian case, also adopting an instrument-based historical perspective on R&I policy design. To conclude, Chap. 6 illustrates the core findings of the analysis. First, it wraps up the historical analysis of French and Italian diachronic R&I instrument evolution. Therefore, it digs into the analysis of national policy mix evolution, testing the relevance of proposed policy instrument classifications. Then, it comparatively analyzes the distinctive impact of party politics' orientation, institutional layouts, and policy takers' responsiveness across the two cases. Finally, it presents the different actors' interaction strategies underpinning the evolution of R&I policy instrument choices across the two cases, while discussing policy implications and related issues for future research.

Bibliography

Amable, Bruno, and Pascal Petit. 2001. The Diversity of Social Systems of Innovation and Production During the 1990s.

Arnold, Erik, and Patries Boekholt. 2003. Research and Innovation Governance in Eight Countries: A Meta-Analysis of Work Funded by EZ (Netherlands) and RCN (Norway).

Béland, Daniel. 2009. Ideas, Institutions, and Policy Change. *Journal of European Public Policy* 16 (5): 701–718. http://www.tandfonline.com/doi/abs/10.1080/13501760902983382 (December 2, 2019).

Ben-David, J. 1991. Scientific Growth. In *Scientific Growth: Essays on the Social Organization and Ethos of Science*, California Studies in the History of Science, 1st ed. University of California Press.

Bergek, Anna, et al. 2015. Technological Innovation Systems in Contexts: Conceptualizing Contextual Structures and Interaction Dynamics. *Environmental Innovation and Societal Transitions* 16: 51–64.

Bhattacharya, Utpal, Po-Hsuan Hsu, Xuan Tian, and Xu. Yan. 2017. What Affects Innovation More: Policy or Policy Uncertainty? *Journal of Financial and Quantitative Analysis* 52 (5): 1869–1901.

Binder, Sarah A., et al. 2009. Elaborating the 'New Institutionalism.'

Borrás, Susana. 2009. The Widening and Deepening of Innovation Policy: What Conditions Provide for Effective Governance? *Papers in Innovation Studies 2009/2, Lund University, CIRCLE –Center for Innovation, Research and Competences in the Learning Economy.*

Borrás, Susana, and Charles Edquist. 2019. *Holistic Innovation Policy.* Oxford University Press. http://www.oxfordscholarship.com/view/10.1093/oso/9780198809807.001.0001/oso-9780198809807 (June 10, 2019).

Bouwma, Irene, Bas Arts, and Duncan Liefferink. 2017. Cause, Catalyst or Conjunction? The Influence of the Habitats Directive on Policy Instrument Choice in Member States. *Journal of Environmental Planning and Management* 60 (6): 977–996. https://www.tandfonline.com/doi/full/10.1080/09640568.2016.1189819 (June 10, 2019).

Braun, Dietmar. 1993. Who Governs Intermediary Agencies? Principal-Agent Relations in Research Policy-Making. *Journal of Public Policy* 13 (02): 135. http://www.journals.cambridge.org/abstract_S0143814X00000994 (September 18, 2018).

———. 2008a. Lessons on the Political Coordination of Knowledge and Innovation Policies. *Science and Public Policy* 35 (4): 289–298.

———. 2008b. Organising the Political Coordination of Knowledge and Innovation Policies. *Science and Public Policy* 35 (4): 227–239.

Bressers, Hans, and Laurence O'Toole. 1998. The Selection of Policy Instruments: A Network-Based Perspective. *Journal of Public Policy* 18: 213–239.

Bressers, H.A., and L.J. O'Toole. 2005. Instrument Selection and Implementation in a Networked Context. In *Designing Government: From Instruments to Governance*, ed. P. Eliadis, M. Hill, and M. Howlett, 132–153. Montreal: McGill-Queens University Press.

Bush, Vannevar. 1945. *Science: The Endless Frontier A Report to the President By Vannevar Bush, Director Fo the Office of Scientific Research and Development.*

Candel, Jeroen J.L., and Robbert Biesbroek. 2016. Toward a Processual Understanding of Policy Integration. *Policy Sciences* 49 (3): 211–231. https://doi.org/10.1007/s11077-016-9248-y.

Cantner, Uwe, and Andreas Pyka. 2001. Classifying Technology Policy from an Evolutionary Perspective. *Research Policy* 30 (5): 759–775. https://www.sciencedirect.com/science/article/pii/S0048733300001049 (October 24, 2019).

Capano, Giliberto, and Andrea Lippi. 2017. How Policy Instruments Are Chosen: Patterns of Decision Makers' Choices. *Policy Sciences* 50 (2): 269–293. https://doi.org/10.1007/s11077-016-9267-8.

Capano, Giliberto, Andrea Pritoni, and Giulia Vicentini. 2019. Do Policy Instruments Matter? Governments' Choice of Policy Mix and Higher Education Performance in Western Europe. *Journal of Public Policy*: 1–27.

Cashore, Benjamin, and Michael Howlett. 2007. Punctuating Which Equilibrium? Understanding Thermostatic Policy Dynamics in Pacific Northwest Forestry. *American Journal of Political Science* 51 (3): 532–551. http://doi.wiley.com/10.1111/j.1540-5907.2007.00266.x (June 11, 2019).

Chaminade, Cristina, and Charles Edquist. 2006. From Theory to Practice: The Use of Systems of Innovation Approach in Innovation Policy. In 141–62.

Chesnais, François. 1993. The French National System of Innovation. In *National Innovation Systems: A Comparative Analysis*, ed. R.R. Nelson, 192–229. New York: Oxford University Press.

Chou, Meng-Hsuan, Jens Jungblut, Pauline Ravinet, and Martina Vukasovic. 2017. Higher Education Governance and Policy: An Introduction to Multi-Issue, Multi-Level and Multi-Actor Dynamics. *Policy and Society* 36 (1): 1–15. https://www.tandfonline.com/doi/full/10.1080/14494035.2017.1287999 (October 24, 2019).

Christensen, Tom, Anne Lise Fimreite, and Per Lægreid. 2014. Joined-Up Government for Welfare Administration Reform in Norway. *Public Organization Review* 14 (4): 439–456.

Chung, Chao-chen. 2013. Government, Policy-Making and the Development of Innovation System: The Cases of Taiwanese Pharmaceutical Biotechnology Policies (2000–2008). *Research Policy* 42 (5): 1053–1071. https://www.sciencedirect.com/science/article/pii/S0048733313000127 (April 29, 2019).

Cimoli, Mario, Giovanni Dosi, and Joseph E. Stiglitz. 2009. *Industrial Policy and Development*. Oxford University Press. http://www.oxfordscholarship.com/view/10.1093/acprof:oso/9780199235261.001.0001/acprof-9780199235261 (October 24, 2019).

Considine, Mark, Jenny M. Lewis, and Damon Alexander. 2009. *Networks, Innovation and Public Policy: Politicians, Bureaucrats and the Pathways to Change inside Government*. Basingstoke: Palgrave Macmillan.

Curley, Cali, Richard Feiock, and Kewei Xu. 2020. Policy Analysis of Instrument Design: How Policy Design Affects Policy Constituency. *Journal of Comparative Policy Analysis: Research and Practice*. https://www.tandfonline.com/doi/abs/10.1080/13876988.2020.1749517 (January 5, 2021).

Dermont, Clau, Karin Ingold, Lorenz Kammermann, and Isabelle Stadelmann-Steffen. 2017. Bringing the Policy Making Perspective in: A Political Science Approach to Social Acceptance. *Energy Policy* 108: 359–368. https://www.sciencedirect.com/science/article/pii/S0301421517303531 (October 24, 2019).

Edler, Jakob, and Jan Fagerberg. 2017. Innovation Policy: What, Why, and How. *Oxford Review of Economic Policy* 33 (1): 2–23. https://academic.oup.com/oxrep/article/2972712/Innovation (June 12, 2019).

Edler, Jakob, and Stefan Kuhlmann. 2008. Coordination within Fragmentation: Governance in Knowledge Policy in the German Federal System. *Science and Public Policy* 35 (4): 265–276.

Edler, Jakob, Paul Cunningham, A. Gök, and Philip Shapira. 2016. *Handbook of Innovation Policy Impact.* Cheltenham: Edward Elgar Publishing.

Edquist, Charles. 2001a. The Systems of Innovation Approach and Innovation Policy: An Account of the State of the Art. In *DRUID Conference,* Aalborg, 12–15.

———. 2001b. The Systems of Innovation Approach and Innovation Policy: An Account of the State of the Art.

———. 2014. Striving towards a Holistic Innovation Policy in European Countries – But Linearity Still Prevails! *STI Policy Review* 5 (2): 1–19.

Edquist, Charles, and Susana Borrás. 2013. The Choice of Innovation Policy Instruments. *Technological Forecasting and Social Change* 80: 1513–1522.

Edquist, Charles, and Jon Mikel Zabala-Iturriagagoitia. 2012. Public Procurement for Innovation as Mission-Oriented Innovation Policy. *Research Policy* 41 (10): 1757–1769.

Fagerberg, Jan. 2017. Innovation Policy: Rationales, Lessons and Challenges. *Journal of Economic Surveys* 31 (2): 497–512. http://doi.wiley.com/10.1111/joes.12164 (July 23, 2020).

Fagerberg, Jan, David C Mowery, and Jan Fagerberg. 2009. "Innovation: A Guide to the Literature."

Flanagan, Kieron, Elvira Uyarra, and Manuel Laranja. 2011. Reconceptualising the 'Policy Mix' for Innovation. *Research Policy* 40 (5): 702–713.

Foray, D., D.C. Mowery, and R.R. Nelson. 2012. Public R&D and Social Challenges: What Lessons from Mission R&D Programs? *Research Policy* 41 (10): 1697–1702. https://www.sciencedirect.com/science/article/pii/S0048733312002193 (October 24, 2019).

Freeman, Christopher. 1987. *Technology Policy and Economic Performance: Lessons from Japan.* London: Pinter Pub LTD.

Gassler, Helmut, Wolfgang Polt, and Christian Rammer. 2008. Priority Setting in Technology Policy: Historical Developments and Recent Trends. In *Innovation Policy in Europe: Measurement and Strategy*, ed. C. Nauwelaers and R. Wintjes. Edward Elgar, Cheltenham.

Ghazinoory, Sepehr, Maghsoud Amiri, Soroush Ghazinoori, and Parisa Alizadeh. 2019. Designing Innovation Policy Mix: A Multi-Objective Decision-Making Approach. *Economics of Innovation and New Technology* 28 (4): 365–385.

Goetz, Klaus H. 2008. Governance as a Path to Government. *West European Politics* 31 (1–2): 258–279. http://www.tandfonline.com/doi/full/10.1080/01402380701835066 (October 24, 2019).

Griessen, Thomas, and Dietmar Braun. 2008. The Political Coordination of Knowledge and Innovation Policies in Switzerland. 35: 277–288.

Gross, Catherine. 2007. Community Perspectives of Wind Energy in Australia: The Application of a Justice and Community Fairness Framework to Increase Social Acceptance. *Energy Policy* 35 (5): 2727–2736. https://www.sciencedirect.com/science/article/pii/S0301421506004861 (October 24, 2019).

Guston, David H. 1996. Principal-Agent Theory and the Structure of Science Policy. *Science and Public Policy* 23 (4): 229–240. https://doi.org/10.1093/spp/23.4.229.

Head, Brian W., and John Alford. 2013. Wicked Problems: Implications for Public Policy and Management. *Administration & Society* 47 (6): 711–739. https://doi.org/10.1177/0095399713481601.

Ingold, Karin, Isabelle Stadelmann-Steffen, and Lorenz Kammermann. 2018. The Acceptance of Instruments in Instrument Mix Situations: Citizens' Perspective on Swiss Energy Transition. *Research Policy*. http://www.sciencedirect.com/science/article/pii/S0048733318302506

Jochim, Ashley E., and Peter J. May. 2010. Beyond Subsystems: Policy Regimes and Governance. *Policy Studies Journal* 38 (2): 303–327. http://doi.wiley.com/10.1111/j.1541-0072.2010.00363.x (April 29, 2019).

Kammermann, Lorenz, and Karin Ingold. 2019. Going beyond Technocratic and Democratic Principles: Stakeholder Acceptance of Instruments in Swiss Energy Policy. *Policy Sciences* 52 (1): 43–65.

Kattel, Rainer, and Mariana Mazzucato. 2018. Mission-Oriented Innovation Policy and Dynamic Capabilities in the Public Sector. *Industrial and Corporate Change* 27 (5): 787–801. https://doi.org/10.1093/icc/dty032.

Kauder, B., and N. Potrafke. 2013. Government Ideology and Tuition Fee Policy: Evidence from the German States. *CESifo Economic Studies* 59 (4): 628–649.

Koch, Christian. 2008. The Superministry Approach: Integrated Governance of Science, Technology and Innovation with Contracted Autonomy. *Science and Public Policy* 35 (4): 253–264.

Laranja, Manuel, Elvira Uyarra, and Kieron Flanagan. 2008. Policies for Science, Technology and Innovation: Translating Rationales into Regional Policies in a Multi-Level Setting. *Research Policy* 37 (5): 823–835.

Laver, Michael, and Kenneth Shepsle. 1996. The Social Context of Government Formation. In *Making and Breaking Governments. Cabinets and Legislatures in Parliamentary Democracies*, ed. Michael Laver and Kenneth Shepsle, 18–42. Cambridge: Cambridge University Press.

Linder, Stephen, and Guy Peters. 1998. The Study of Policy Instruments: Four Schools of Thought. In *Public Policy Instruments: Evaluating the Tools of Public Administration*, ed. B. Guy Peters and Frans K.M. van Nispen. Cheltenham: Edward Elgar Press.

Lundvall, B.A. 1992. *National Systems of Innovation: Towards a Theory of Innovation and Interact*. London: Pinter.

Ma, Liang. 2017. Political Ideology, Social Capital, and Government Innovativeness: Evidence from the US States. *Public Management Review* 19 (2): 114–133.

Majone, Giandomenico. 1976. Choice among Policy Instruments for Pollution Control. *Policy Analysis* 2: 589–613.

Mazzucato, Mariana. 2013. *The Entrepreneurial State: Debunking Public vs. Private Sector Myths*. London: Anthem Press.

Meissner, Dirk, and Sandrine Kergroach. 2019. Innovation Policy Mix: Mapping and Measurement. *The Journal of Technology Transfer* 46: 197.

Merton, R.K. 1938. Social Structure and Anomie. *American Sociological Review* 3 (5): 672–682.

Nauwelaers, Claire, and René Wintjes. 2008. *Innovation Policy in Europe: Measurement and Strategy*. Cheltenham: Edward Elgar.

Neicu, Daniel, Peter Teirlinck, and Stijn Kelchtermans. 2016. Dipping in the Policy Mix: Do R&D Subsidies Foster Behavioral Additionality Effects of R&D Tax Credits? *Economics of Innovation and New Technology* 25 (3): 218–239.

Nelson, R. 1993. *National Innovation System: A Comparative Analysis*. Oxford: Oxford University Press.

Nilsson, Måns, and Lars Nilsson. 2005. Towards Climate Policy Integration in the EU: Evolving Dilemmas and Opportunities. *Climate Policy* 5: 363–376.

OECD. 2005. *Governance of Innovation Systems*. OECD.

———. 2014. *Reviews of Innovation Policy: France 2014*. OECD.

———. 2015a. *Frascati Manual 2015*. OECD.

———. 2015b. *The Innovation Imperative*. OECD.

OECD/Eurostat. 2018. *Oslo Manual 2018: Guidelines for Collecting, Reporting and Using Data on Innovation, 4th Education*, The Measurement of Scientific, Technological and Innovation Activities. Paris: Eurostat.

Palier, Bruno. 2007. Tracking the Evolution of a Single Instrument Can Reveal Profound Changes: The Case of Funded Pensions in France. *Governance* 20 (1): 85–107.

Pelkonen, Antti, Tuula Teräväinen, and Suvi-Tuuli Waltari. 2008. Assessing Policy Coordination Capacity: Higher Education, Science and Technology Policies in Finland. *Science and Public Policy* 35 (4): 241–252.

Pencheva, Irina, Marc Esteve, and Slava Jankin Mikhaylov. 2018. Big Data and AI – A Transformational Shift for Government: So, What Next for Research? *Public Policy and Administration* 35 (1): 22–44.

Peters, B. Guy. 1998. *Managing Horizontal Government: The Politics of Coordination*. Ottawa: Canadian Centre for Management Development.

———. 2017. What Is so Wicked About Wicked Problems? A Conceptual Analysis and a Research Program. *Policy and Society* 36 (3): 385–396.

Ragin, Charles C., and Howard Saul Becker. 2009. *What Is a Case? Exploring the Foundations of Social Inquiry*. Cambridge University Press.

Ringeling, Arthur B. 2005. Instruments in Four: The Elements of Policy Design. In *Designing Government: From Instruments to Governance*, ed. P. Eliadis, M.M. Hills, and M. Howlett, 185–202. Kingston: Ontario McGill-Queens University Press.

Salamon, Lester. 2002. The New Governance and the Tools of Public Action. *Fordham Urban Law Journal* 28 (5).

Sanz-Mendez, Luis, and Susana Borrás. 2001. Explaining Changes and Continuity in EU Technology Policy: The Politics of Ideas. In *Changing European Research System*, ed. Imon Dresner and Nigel Gilbert. Aldershot: Ashgate Press.

Scharpf, Fritz Wilhelm. 1997. *Games Real Actors Play: Actor-Centered Institutionalism in Policy Research*. Boulder: Westview Press.

Schneider, Anne, and Helen Ingram. 1993. Social Construction of Target Populations: Implications for Politics and Policy. *American Political Science Review* 87 (2): 334–347.

Scholten, Peter, Elizabeth Collett, and Milica Petrovic. 2017. Mainstreaming Migrant Integration? A Critical Analysis of a New Trend in Integration Governance. *International Review of Administrative Sciences* 83 (2): 283–302.

Smits, Ruud, Philip Shapira, and Stefan Kuhlmann. 2010. *The Theory and Practice of Innovation Policy: An International Research Handbook*. Cheltenham: Edward Elgar.

Stoppino, Mario. 2001. *Potere e Teoria Politica*. Giuffre.

Tosun, Jale. 2018. Investigating Ministry Names for Comparative Policy Analysis: Lessons from Energy Governance. *Journal of Comparative Policy Analysis: Research and Practice* 20 (3): 324–335.

Tosun, Jale, and Achim Lang. 2017. Policy Integration: Mapping the Different Concepts. *Policy Studies* 38 (6): 553–570.

Trein, Philipp. 2017. A New Way to Compare Horizontal Connections of Policy Sectors: 'Coupling' of Actors, Institutions and Policies. *Journal of Comparative Policy Analysis: Research and Practice* 19 (5): 419–434.

Ulnicane, Inga. 2016. 'Grand Challenges' Concept: A Return of the 'Big Ideas' in Science, Technology and Innovation Policy? *International Journal of Foresight and Innovation Policy* 11 (1/2/3): 5.

Van Asselt, H., T. Rayner, and Å. Persson. 2015. Climate Policy Integration. In *Research Handbook on Climate Governance*, ed. K. Bäckstrand and E. Lövbrand, 388–399. Cheltenham: Edward Elgar.

Van de Ven, A.H., and R. Garud. 1989. A Framework for Understanding the Emergence of New Industries. In *Research on Technological Innovation and Management Policy*, ed. R. Rosenbloom and R. Burgelman. Greenwich: JAI Press.

Varone, Frédéric, and Bernard Aebischer. 2001. Energy Efficiency: The Challenges of Policy Design. *Energy Policy* 29 (8): 615–629.

Verhoest, Koen, and Geert Bouckaert. 2005. Machinery of Government and Policy Capacity. In *Challenges to State Policy Capacity*, ed. M. Painter and J. Pierre. Basingstoke: Palgrave.

Wang, Quan-Jing, et al. 2019. The Impacts of Government Ideology on Innovation: What Are the Main Implications? *Research Policy* 48 (5): 1232–1247.

WEF. 2020. *Global Risks 2020, 15th Edition*. Geneva: World Economic Forum.

Wilson, Carter. 2000. Policy Regimes and Policy Change. *Journal of Public Policy* 20 (3): 247–274.

Woodside, Kenneth. 1986. Policy Instruments and the Study of Public Policy. *Canadian Journal of Political Science/Revue canadienne de science politique* 19 (4): 775–793.

CHAPTER 2

The Determinants of Policy Design Choices: A Theoretical Framework for Analysis

As discussed in Chap. 1, Research and Innovation (R&I) represents a case of complex governance arrangements where policymakers are often bounded to share their authority with a plurality of actors. Hence, the more the power is dispersed, the more different actors will have a say in the decision-making process. Consequently, it becomes more difficult to foster the necessary consensus to support changes. This sector, as in many other fields characterized by integrated policy issues, experiences a reduced capacity of political institutions to act unilaterally, and a consequently high degree of resource dependency between target population and policymakers. In this context, policymaking requires both internal (when coordinating different institutions) and external (when interacting with policy recipients) governance ability. Consequently, to understand how R&I policies are framed and put into practice, a greater attention toward the actors involved along this process is needed.

Policymaking is a complex political and technical process influenced by the nature of the problems at hand and by previous experiences of tackling similar issues (Howlett and Ramesh 1993). However, also the subjective preferences of decision-makers and the likely reaction of the social groups affected by these choices have been found to influence this process (Borrás and Edler 2015; Capano and Lippi 2017; Edler et al. 2016; Edler and Fagerberg 2017; John 2012; Laranja et al. 2008; Linder and Peters 1989; Ringeling 2005). Investigating these activities means

C. Acciai, *Policy Design for Research and Innovation*, International Series on Public Policy,
https://doi.org/10.1007/978-3-031-36628-4_2

considering the political priorities of those actors involved, the power and the stakes of the target population, and the political and administrative costs associated with different alternatives (Rist 1998). This chapter aims at introducing the reader to the conceptual toolkits necessary to break-down this complexity. It starts by introducing the core concepts of policy design analysis and it moves on with a discussion on the scholarship contributions to the sector. Finally, the analytical framework adopted for de-coupling policy design in complex governance regimes is presented and discussed.

2.1 Policy Design: An Overview of the Literature and Core Concepts

Policy design is a branch of the policy sciences pertaining to the study of policy processes. It investigates the ways public policies are constructed from governing resources to achieve results (Howlett and Mukherjee 2018). Transforming policy ambitions into practice is a complex process that lays its foundations in the combination of different techniques of governance used to achieve policymakers' goals: these are called policy instruments. Therefore, if on the one hand policy design is the process during which policy ambitions are transformed into practices, policy instruments substantiate the means to achieve those goals. Instruments are often combined into mixes both because of the increasingly complex policy objectives they aim at solving (e.g., climate change mitigation strategies require different types of "actions," hence possibly multiple types of instruments) and because of the diachronic stratification of policy choices across governments and organizations (e.g., a new cabinet in office chooses to implement new instruments, while the previous cabinet's policies are running).

The scholarship has discussed the way tools interact with each other, the coherence and contradictions of blending instruments into mixes, and the way different combinations of tools influence public action (Howlett and Rayner 2013; Rogge and Reichardt 2016; Schmidt and Sewerin 2019). This is happening because policy instruments trigger behavioral changes in recipients, which consequently alter their interactions with others and their context. However, despite policy instruments representing attempts to "get people to do things they might not otherwise do" (Schneider and Ingram 1993:513), policymakers are also part of a broader decision-making environment, where they are held accountable for how they exercise their powers (Considine and Afzal 2010).[1] On the one hand, it is unquestionable how policy instruments represent devices to "get things done." They alter power dynamics (Lascoumes and Le Gales 2007; Salamon 2011) and influence the framing of policies (Baumgartner and Jones 1993), ultimately introducing opportunities and constraints on their targets (Schneider and Ingram 1990). However, on the other hand, democratic governments are not free in selecting the instruments they prefer, "only authoritarian states do not have to be bothered by these limitations" (Ringeling 2005:199). On top of that, in integrated policy sectors, like R&I policies, the burden of design choice is increasingly shared across different policymakers and stakeholders creating new configurations of policymaking institutions (Borrás and Edquist 2019). Consequently, the increased complexity of policy problems limits the capacity of political institutions to act unilaterally and this has consistently modified decision-making settings (Bressers and O'Toole 2005; Capano and Lippi 2017; Majone 1989; Ringeling 2005; Woodside 1986). This means that state structures are evolving both internally, with a reorganization of administrative structures, and externally in their relationship with citizens and the market. Crucial elements of public authority are shared with a multitude of non-governmental actors, who are also involved in the management and implementation of public policies. Therefore, to understand the underlining forces influencing R&I policy design choices, it is necessary to investigate this process from a two-way perspective, where both policymakers and policy takers have the power to influence the likelihood of certain instruments' features to be selected.

In the policy-instrument literature we find support for the hypothesis that the resources needed to activate policy instruments are dependent upon the participation of different actors and the assets they can mobilize (Béland and Howlett 2016; Capano and Lippi 2017; Flanagan et al. 2011; Majone 1976). However, we still don't know much about the way the interactions, and especially the interdependency between policymakers and policy takers, influences R&I policy design choices. Table 2.1 summarizes the most relevant scholarly contributions on these topics. These are classified according to the main factors influencing policy-instrument selection, the role of target population, and the main theoretical assumptions characterizing each perspective.

[1] This is intended as the formal rules and procedures regulating actors' interaction.

Table 2.1 Policy design literature: a focus on instruments

	Key variables for selection process	*Target population role*	*Assumptions*
Wilson (1973)	Governments prefer to start by first adopting the least coercive instruments, while moving up the scale (When necessary, to overcome technical resistance to effective regulation)	Recalcitrance/social pressure by target groups drive toward higher coercive instruments	All instruments are technically substitutable In liberalist governments, as opposed to interventionists, we should find more efforts to maintain instruments at a low level of coercion
Majone (1976)	It is dependent upon the relative distribution of power, political constraints, skills of policymakers to exploit the degrees of freedom allowed by the system	The multiplicity of players constantly attempts to modify the rules of the game and to create powerful oppositions to prevent their selection	Regulators, regulated and other interest groups interact in a single decisional structure
Trebilcock et al. (1982)	The interaction between constraints on government and the extent to which an instrument can be substituted with another help to explain their choice	Decision-makers will select instruments that confer benefits on marginal voters and costs on infra-marginal ones. This will be done in a way that accentuates their visibility while minimizing their costs	Full rationality of decision-makers Voters' ignorance about the influence of policy output
Woodside (1986)	The lower the political power of the target population, the higher the likelihood of adopting highly coercive policy instruments. This probability decreases where the more powerful actors are	The nature of the target group represents a source of differentiation in instruments' features. Because certain groups will be able to demand more respect and autonomy from state officials for their interests	Both the choice and the way in which an instrument is structured will reflect the political power of the policy clientele

(*continued*)

Table 2.1 (continued)

	Key variables for selection process	*Target population role*	*Assumptions*
Hood (1986)	Instrument choice is a function of the nature of state goals and resources, together with organizational capacity of target groups	Different instruments vary in their effectiveness according to the nature of the social group they are intended to influence. The larger the target group (and the better organized) the more likely it is that governments will use passive instruments	Capability of decision-makers to be united and overcome collective action problems.
Linder and Guy Peters 1989	Organizational culture of the agency: the context of the problem situation and decision-making, subjective preferences for instruments' combinations understood as being socially acceptable	The relevance of contextual factors and the social acceptability of instruments reflect target groups' relevance (the nature of the clientele served by the organization)	Some instruments are more apt than others according to the context. The attributes of any given instrument are thought to matter less than the prevailing conditions
Schneider and Ingram (1993)	The allocation of benefits and burdens embedded in the action of different policy instruments varies to the extent of power (strong vs. weak) and the type of social construction (positive vs. negative) of target population	Social construction of target population[a] influences the agenda and the selection of policy tools as well as the rationalities that legitimize policies. Because different target populations receive different messages embedded in each policy	Target populations are assumed to have empirically verifiable boundaries (because policies create those) and to exist within objective conditions even though those conditions are subject to multiple evaluations (p. 335)

(*continued*)

Table 2.1 (continued)

	Key variables for selection process	*Target population role*	*Assumptions*
	Key variables for selection process	**Target population role**	**Assumptions**
Van der Doelen (1998)	By combining stimulative (low coercion) and repressive (high coercion) forms of control models, policymakers enhance the feasibility and effectiveness of a policy	Legitimacy: the degree to which a certain instrument is assessed as feasible by policymakers and evokes acceptance among citizens and organizations Effectiveness: the degree to which the chosen policy instrument contributes to goals' attainment	Policy instruments have two purposes: to legitimate a policy and to effectuate it
Howlett (2000)	The selection varies according to the type of instrument Substantive: it is dependent on state capacity and complexity of social actors' state wishes to influence Procedural: It is dependent on the interplay between sectorial de-legitimation and level of systemic de-legitimation	Their characteristics and the extent of trust and legitimacy they require influence the type of instrument selected	Substantive instruments: are intended to directly affect the nature, types, quantities, and distribution of the goods and services provided by the society Procedural instruments: have the purpose to alter or manipulate the policy process (used to manipulate number/ nature of actors arrayed in policy subsystems)
Macdonald (2001)	The balance of power between regulators regulated	Target population attitudes influence the coerciveness associated with the political dynamics at the governmental level and variations in the level of instruments' enforcement	The two main actors of a network will always function as a focal point for a larger network

(*continued*)

Table 2.1 (continued)

	Key variables for selection process	*Target population role*	*Assumptions*
Howlett and Ramesh (2003)	The nature of the government in question and social actors' behavior that the government is wishing to influence	The likely reaction to the choice of the instrument by target groups influences the selection process	The choice is shaped by instruments' characteristics, the nature of the problem, and governments' past experiences in dealing with the same/similar problems
Bressers and O'Toole (1998,2005)	Network interconnectedness and cohesion[b] influence the selection process	Government actions are not simply and unilaterally authoritative; they depend on the action and the acquiescence or support of others	The course and outcomes of the process depend on inputs, but also on the characteristic of the actors involved (their motivations, information, and relative power)
Capano and Lippi (2017)	Different patterns of selection based on legitimacy (sense making) and instrumentality (search for effectiveness) required for their selection	The instrument must be accepted by some (the public, the specific target groups, experts or external actors) in terms of the values of that specific sector or the instruments' capacity to achieve certain goals	Decision-makers are rational beings trying to find good reasons for believing that a policy instrument is suitable and useful

[a]These are cultural characteristics or popular images of the persons or groups whose behavior and well-being are affected by public policy (Schneider and Ingram 1993:334)

[b]Interconnectedness refers both to the contacts in the relevant policy formation process (and the habits that have developed in this connection over time) and also to the relationship between these actors outside the actual policy process at any particular time. Meanwhile, cohesion refers to the extent to which individuals, groups, and organizations emphasize with each other's objective insofar as these are relevant in other policy fields. This empathy generally stems from shared values and shared worldview (Bressers and O'Toole 1998:219)

The literature highlights how the interplay between state preferences, organizational capacity, and the willingness to cooperate between stakeholders are among the most relevant factors shaping policy-instrument choices. This is an inherently political process involving different actors, interests, and constraints (Linder and Guy Peters 1989; Peters 2000) and it rather resembles a ***muddling through*** (Lindblom 1959). This suggests how policy design is actually influenced by power dynamics, by the political, institutional, and social contexts in which decisions take place, and by the availability of state resources as well as by the preferences of target population (Bressers and O'Toole 1998; Hood 1986; Linder and Guy Peters 1989; Macdonald 2001; Majone 1976; Schneider and Ingram 1993; Trebilcock and Hartle 1982). In addition, different national contexts and public administration capacities can influence this process too (Borrás and Edquist 2019). Therefore, a deeper understanding of the context in which instruments are selected, of the actors involved and their organizational structure, is needed for comprehending the drivers of these choices. The innovative contribution of this study is namely to investigate policy design from the perspective of both the actors "making" and the actors "receiving" those decisions. However, before delving into the analytical approach adopted for analyzing this relationship, we will take a step back and introduce the outcome variable of this research: policy instruments.

2.2 The Outcomes of Policy Design: Policy Instruments and Mixes

Policy instruments are different techniques of governance by means of which policymakers connect their preferences with the formulation of different strategies in the effort to achieve their goals. Policy design refers to the process throughout which actors select and model instruments, while policy instruments themselves are the basic unit of thinking when investigating the actions of government. Moving from the single instrument to their combination means focusing on policy mixes (Rogge 2018). These are blends of instruments expected to address more complex problems that cannot be achieved by means of a unique instrument. Within this jargon it is important to differentiate between policy mixes—adopting a diachronic perspective to indicate the overtime stratification of policy tools—and instrument mixes, which refer to a temporally defined set of instruments and their configurations. Policy mixes imply a longitudinal perspective of analysis, which diachronically follows the evolution of instrument blends. Differently, instrument mixes indicate a cross-sectional view, focusing on a temporally defined "screenshot" of the instrument blend at a specific point in time.

The literature on policy instruments is vast and it spans across different policy areas. This scholarship has produced a plurality of typologies indicating "ways of doing different kinds of analysis, rather than different ways of doing the same kind of analysis" (Hood and Magretts 2007:141). This results in alternative instrument classifications that are not interchangeable because they assign different meanings to what instruments are (Acciai and Capano 2021). For example, according to Vedung (1998), policy instruments represent "a set of techniques by which governmental authorities wield their power, while attempting to ensure support and affect social changes" (p. 3). These are ends in themselves, because of their capacity to influence the policymaking process. Indeed, Lascoumes and Le Galés (2007) define instruments as institutions, since they can influence the way actors behave, driving forward representations of problems, while organizing social relations between policymakers and takers. Similarly, Salamon (2002, 2011) claims that "tools significantly structure decision-making networks, by defining actors that are centrally involved in a particular type of program and the formal rules they will play" (p. 64). Others focus on the role of policy instruments as techniques and devices governments use to implement policies and achieve their goals (Howlett et al. 2009; Schneider and Ingram 1990), and define instruments as means through which governments attempt to shape the life of its constituents (Hood 1986) and make use of legitimate coercion (Doern and Phidd 1983). Table 2.2[2] summarizes the most relevant developments in policy-instrument literature while illustrating the theoretical foundations behind their definition and classification.

The policy-instrument treatment adopted for the current analysis builds upon an enriched variation of Vedung's (1998) policy instruments typology. With analytical simplicity (Salamon 2002), this classification differentiates instruments according to "how much events will be controlled and regulated by government authority, and how much they will be left up to the voluntary initiatives of individuals" (Vedung 1998:22). Moreover, by focusing on the drivers of expected behaviors, this theorization overcomes

[2] Adapted from (Acciai and Capano 2021).

Table 2.2 Policy instrument classifications

	Doern and Phidd (1983)	*Hood (2007, 1986)*	*Schneider Ingram (1990)*	*Vedung (1998)*	*Howlett (2000)*	*Salamon (2002, 2011)*	*Lascoumes and Le Galés (2007)*
Meaning of policy instruments	Means for the management and manipulation of legitimate coercion	Means through which governments attempt to shape the life of their constituents. They are about social control	Techniques addressing the problem that people are not taking actions needed to ameliorate social, economic, or political problems	Set of techniques by which governmental authorities wield their power attempting to ensure support and affect social change. They are ends in themselves as they represent the content of the political debate	Techniques and devices governments use to implement policies and achieve their goals	Identifiable methods through which collective action is structured to address a public problem They define the actors centrally involved and the formal role they will play	Devices that are both technical and social, that organize specific social relations between the state and those it is addressed to, according to the representations and meanings it carries

Classification	Based on the extent of coercion adopted for instrument action	NATO model resources, at governmental level simply by virtue of being governments (1983:4). Nodality, authority, treasure, organization[a]	Different assumptions about how policy-relevant behavior can be fostered by the target population by providing: legitimation of authority; tangible payoffs (positive or negative); resources to enable individuals to make decisions; alternative values and beliefs (persuasion)	Inspiration from Etzioni (1961) according to the type of relationship between governors and governed	NATO model inspired by Hood (1986)	Because of the multidimensionality of policy instruments, no single classification is possible, and these schemes will differ depending on which facet is used as the basis[b]	Have classified five major typologies of tools relying partly on the NATO model, improving, and supplementing it by adding the type of political relations they create once in force and the legitimacy required to be applied

(*continued*)

Table 2.2 (continued)

	Doern and Phidd (1983)	*Hood (2007, 1986)*	*Schneider Ingram (1990)*	*Vedung (1998)*	*Howlett (2000)*	*Salamon (2002, 2011)*	*Lascoumes and Le Galès (2007)*
Tool dimensions	Visibility, historical context, politicians prefer less coercive instruments	Defective instruments (used to gather information). Effectives (used to modify behaviors)	Different responses to the lack of action by target population: authority tools, incentive tools, capacity tools, symbolic and hortatory tools, learning tools	Action content: what the target population should do. Authoritative force: degree of power needed to obtain compliance	What they do "substantive" (modify distribution of goods and services); where do they operate "procedural" (indirectly influence policy outcomes through the manipulation of policy process)	Type of good delivery; delivery vehicle, the set of organizations designated to the provision of the service and a system of rules	Related to their effect: the type of political relations organized by instruments and the type of legitimacy that such relations presuppose

[a]Nodality: the capacity of the government to operate as a node in information networks; authority denotes government's assets or fungible resources; organization denotes its capacity for direct action

[b]The author provides a range of dimensions according to which it is possible to compare different sets of tools, namely: coerciveness (the degree of authoritative power instruments rely on), directness (the degree of involvement of the authorizing and financing actor in the precision of the service), automaticity (the extent to which instruments make use of existing administrative structures), and visibility (the extent to which resources devoted to a toll are visible in the governing budget). He believes that it is the only way through which it is possible to clarify the full matrix of choice policymakers face and the trade-offs between them (Salamon 2011)

the strictly resource-based approach of other classifications (Hood 1986; Hood and Magretts 2007; Howlett 2000). Such a variation is very important because resources needed for governing are not only localized at the governmental level, "simply by virtue of being governments" (Hood 1986:4), but these are rather dependent upon the action and acquiescence or support of others, which cannot always be directly controlled (Bressers et al. 1998).

Since public policy attempts to "get people do things they might not otherwise do" (Schneider and Ingram 1990:513), this research adopts an operationalization of tool dimensions focusing on the drivers of expected behaviors induced by policy instruments according to three hierarchical components:

- Authority: describes the degree to which governments use their power in order to induce specific behaviors on target population (Capano et al. 2019); by illustrating the means-end rationalities embedded in instrument choices.
- Instrument shape: specifies the characteristics of the substantive requirements of behaviors (inducement) administered to the target population. This varies according to the degree of coercion they apply on the target population to obtain compliance or to deliver expected outcomes.
- Delivery: represents the types of governing arrangements accompanying instruments, therefore the way tools enforce their action on recipients—that is, activation process—(Salamon 2002, 2011). Because, in addition to social control, every instrument embeds a particular way of exercising it, by privileging some actors and interests over others (Kassim and Le Galès 2010; Lascoumes and Le Gales 2007).

The following paragraphs delve into the analytical characteristics of the proposed policy-instrument typology and the dimensions differentiating their action. Authority, instrument shapes, and delivery components are nested within instrument action. This means that policy instruments can conjugate different combinations of the three components. However, each component is mutually exclusive. Therefore, an instrument cannot be at the same time a regulatory and an economic instrument or a low and a high automatic instrument. This is the reason why the concept of instrument mix has been introduced, to explain how different policy tools are combined to address issues that cannot be appropriately treated with only one instrument. These dimensions are further discussed, and combined with empirical evidence, when analyzing the diachronic evolution of policy instrument choices in France and Italy (Chaps. 4 and 5). Moreover, in the following paragraphs, and in Chap. 3, the problematic of their selection and blending into a mix are presented and discussed considering policy design theories and the analytical framework developed in the current analysis.

2.2.1 *Authority*

Authority represents the common defining feature embedded in any policy instrument and it describes the degree to which governments use their legitimate power to steer policy recipients toward their intended behaviors. Differences among instrument families concern "how much events will be controlled and regulated by government authority, and how much they will be left up to the voluntary initiatives of individuals" (Vedung 1998:22). Families of instruments clarify different typologies of authoritative relationships between decision-makers and recipients of public action, describing the means-end rationality of the inducement chosen. These dimensions are organized on a continuum from the most (regulation)[3] to the least (information) authoritative instrument (Fig. 2.1).

Regulatory instruments implement limitations of certain types of activities backed up by formulated rules and directives (Vedung 1998:31). They require (or forbid) certain behaviors of recipients, under the threat of penalties identified by the rules embedded within them.[4] At the medium level of authority, expenditure instruments demand requirements, or behaviors, in exchange for subsidies (money, time, services, or material benefits) they

[3] Here, regulation is intended as one of the varieties of tools that governments have at their disposal to exert power (Vedung 1998:30–31).

[4] Our interpretation of institutions relates to the explicit norms and rules constraining individual actors' choices and influencing their behavior, meaning the set of shared codes and beliefs influencing the rule of the game (John 2012; North 1990; Peters 2012). The creation of public agency represents a highly coercive instrument. Indeed, such a change drives the modification of the institutional structure and of the system of power relations among the actors involved in a given sector.

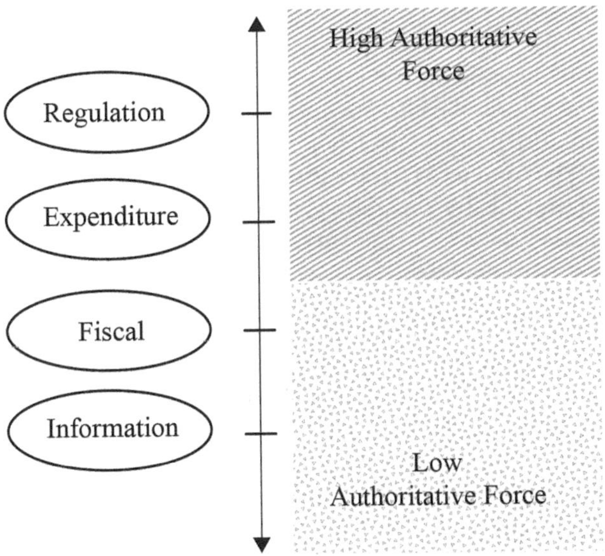

Fig. 2.1 Instrument families

make available. Down in the continuum, fiscal measures neither forbid nor require a given action, but they simply make it costlier/easier to pursue certain behaviors. They do so by affecting the cost of alternative courses of action open to recipients. An important addition to the original classification of Vedung (1998) is the separation between expenditure and fiscal measures (which are clustered inside economic instruments by the former). This exercise is intended to fully grasp the heterogeneity of constraints public action can implement. This is because the power relationship between policymakers and policy recipients will be different in the case of expenditure-oriented economic tools (where governments provide funds to certain actors/activities for a given purpose) versus taxation-oriented economic tools (where governments reward or charge certain behaviors of the target population). Therefore, economic and taxation instruments can be thought as having a different political economy exactly because they can create different behavioral constraints on recipients (Capano et al.

2019; Woodside 1983).[5] These two families differ in the underlying inducement involved in the governing effort, basically to the extent of authority which policymakers apply to steer target populations toward the intended behaviors. Finally, the least authoritative set of instruments are information instruments, which rely on the voluntary cooperation of recipients. They influence target groups' conduct through moral persuasion, by filling the information gap between actual and potential new behaviors.

2.2.2 *Instrument Shapes*

Each family of tools displays a high degree of heterogeneity in how their basic inducement is shaped to obtain compliance from the target population. While instruments can be grouped into families, according to the degree to which governments use their power, these can be applied in different ways (Vedung 1998), each of them embeds a high level of variation in the constraints they can enforce—that is, action content— (Woodside 1986). Consequently, tools can activate different substantive requirements of behaviors, according to the degree of coercion they apply on the target population. This is what Salamon (2002, 2011) defines as instrument shapes; they represent "the basic analytical units to adopt when assessing how policies are made" (Capano et al. 2019:5). An example of the different design features embedded in policy instrument shapes is the case of loans and grants. Both shapes fall under the family of expenditure, and they are both used to encourage specific activities through the provision of funds. However, they differ in the degree of coercion they apply to the target population. Due to the expectation of repayment, and possibly to the fulfillment of some conditions attached to it, loans are more

[5] These two types of instruments create different typologies of preferences and incentives on the behavioral decisions of recipients: "I'll give you money to do this (expenditure)"; "If you do (don't do) this I'll give you money (taxation)." Subsidy involves money that comes to recipients from the government in the form of, for instance, a grant (a transaction that involves government ownership and control over the money prior to receipt of the grant); whereas a fiscal instrument involves money that the government has not taken from the recipient, which suggests that the recipient has been, and remains, the owner of those funds. A second difference is their visibility. While expenditures will be funded through taxpayers' contribution, the tax expenditure involves foregone revenue. The importance of visibility is related to the fact that the exercise of power entails winners and losers, and thus conflict. Therefore, the extent to which the action is easily associated with the government can become an important variable in government decision-making {Formatting Citation}.

coercive than grants, which basically identify a goal (or a behavior) and provide funds for it. This demonstrates how any tool is actually a "package" of different elements that can be disaggregated into relatively small units and scored according to different behavioral dimensions (Schneider and Ingram 1990). Figure 2.2 illustrates a classification of the most employed instrument shapes in R&I according to policy instrument families.[6]

Each shape is organized along a continuum from the most to the least authoritative family, while the within-shape description is listed by decreasing coercive order, meaning according to the extent to which the target population is free to choose alternative behaviors.

2.2.3 *Delivery*

Policy design doesn't terminate with the enactment of policy decisions, but it rather continues along the structure of political relations that instruments unfold (Lascoumes and Le Gales 2007). Establishing an instrument involves a number of different tasks, and each can attract the interests of various actors (Simons and Voß 2018:21). Therefore, instrument delivery components describe how, in addition to social control, every tool embeds a particular way of exercising it, by privileging some actors and interests over others (Kassim and Le Galès 2010). This provides information on the types of governing arrangements coming together with the implementation of each tool, by defining the parties involved in their management (Bouwma et al. 2017).

Policy instruments require different amounts of government involvement in their activation, and they can count on the participation of different stakeholders along this process, comparing structures of control based on preexisting self-operating systems (market-based), to techniques requiring the cooperation of public agencies or executive bureaucracies (Salamon 2002). Consequently, the dimension of automaticity provides a fine-grained contribution on the role of actors engaged in procedures following policy-instrument formulation, and their power to steer their

[6] This list is the result of the analysis of secondary sources, and it will likely be modified in the future because the variety of policy instruments available to decision-makers is limited only by their imagination (Howlett 2009:114). However, it still represents a satisfactory overview of the different instrument shapes of governing action in R&I.

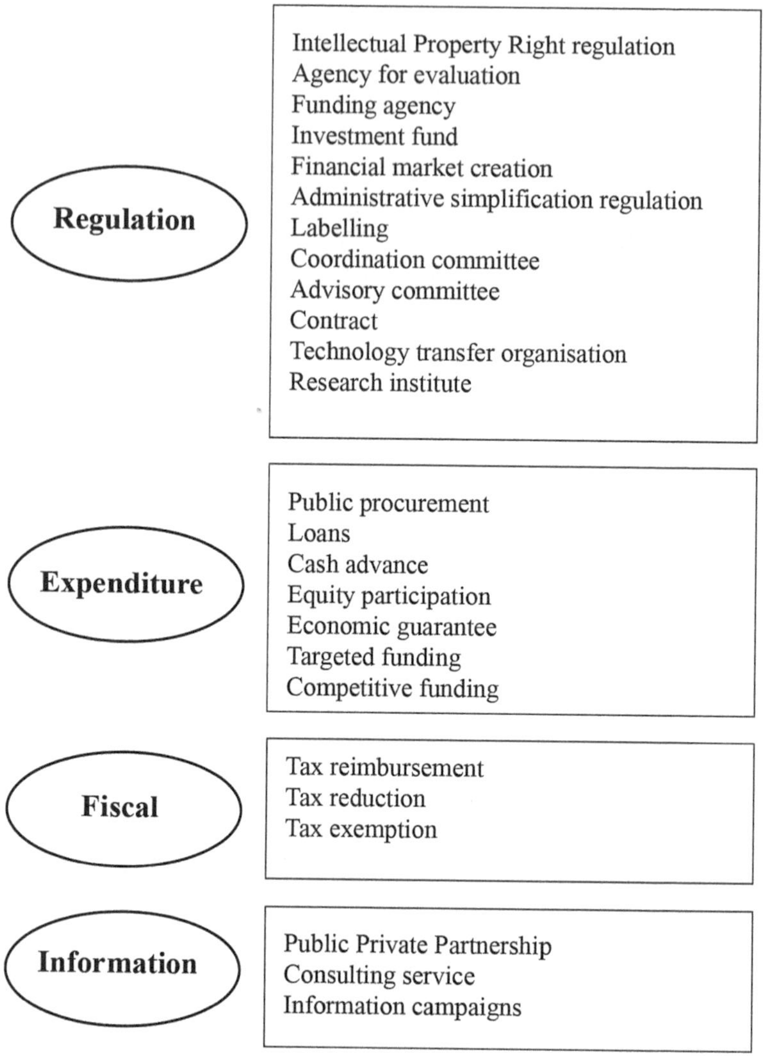

Fig. 2.2 Instrument shapes (by authoritative dimension)

enactment.[7] Figure 2.3 illustrates their classification on a continuum from low to high automaticity. Low automatic instruments rely on administrative or organizational structures (like bureaucracy, public agency, local-regional actors). For example, expenditure instruments—like competitive funds for research—require the involvement of multiple actors drafting the call for funding, collecting, and assessing submissions, managing the provision of funds and the ex-post evaluation procedures. Consequently, in the case of this specific combination of instrument dimensions, the enactment of the instrument relies on setting up an ad-hoc organizational structure. Differently, automatic instruments make use of already-existing structures—like the market, the fiscal or private credit system—hence requiring a different extent of resource investments. An example of this system is tax deductions for investments in R&D for enterprises which do not require the direct involvement of public bureaucracies. Here, recipients can autonomously benefit from the instrument by deducting tax expenses on innovative investments (e.g., advanced machinery, new skills) directly from their balance sheets.

Low and high automatic delivery differ in their visibility and in the political burden attached to their effects. The former tends to be highly visible because of the bureaucratic machinery they put in place for their enactment; high automatic instruments are less visible, since the effective enactment of tools is mainly evident to the recipient and the grantor rather than to the general audience. The mixed type of delivery hinges upon the cooperation with stakeholders populating a specific sector—in this case,

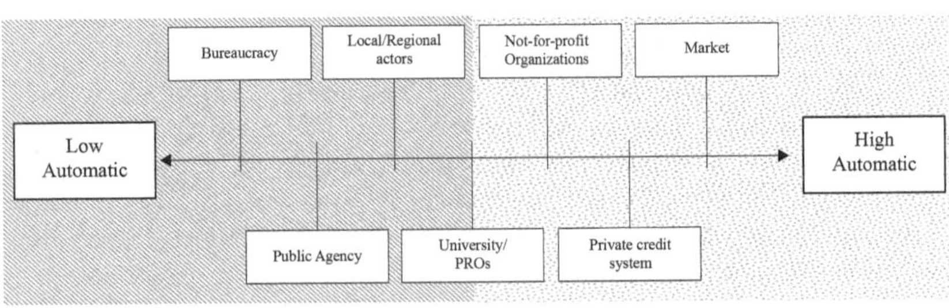

Fig. 2.3 Instrument delivery component

[7] A "certain overlap exists, between the automaticity dimension and the directness dimension of tools. However, not all automatic tools are indirect and not all indirect tools are automatic" (Salamon 2002:1663).

R&I performers like universities, public research organizations, enterprises, venture capitalists, or foundations.

Delivery component is highly context-dependent because the same instrument can assume different degrees of automaticity according to the specific way in which it has been activated. For example, loans (economic family) can be either directly provided by governments (low automatic) or by private lenders like banks (high automatic). Instruments differ on the basis of the amount of public resources necessary for their management, comparing structures of control based on already-existing self-operating systems, to command-and-control techniques which have to be undertaken by public agencies or executive bureaucracies. Therefore, just as the choice of different policy instruments will reflect the political clientele of the policy, the way in which they are structured and assembled may also reflect similar considerations (Woodside 1986).

2.3 Understanding R&I Policy Design: An Explanatory Framework

Decisions regarding policy design are made in a nested context characterized by a complex regime of goals, instruments, and settings in which new choices have to be adjusted within existing contexts (Cashore and Howlett 2007). Thus, the combination of policy instruments selected can vary according to context-specific characteristics, as defined by different configurations of actors, institutional settings, and power struggles. Therefore, to understand the determinants of policy design choices—how and why specific combinations of R&I instrument mixes have been assembled—it is necessary to specify these circumstances. The explanatory framework of this research provides a lens for their interpretation. It intersects insights on the role of politics (as the influence of dominant framings on the types of policies implemented), the opportunity structures provided by the institutional system (internal coordination and specialization between bureaucrats), and the role of the target population (the strategies for interest intermediation) to understand policy design choices. As discussed in the review of the literature (Table 2.1) and in Chap. 1, these factors play an important role in shaping policy design. The first conceptual pillar focuses on the formulation of policy goals. It zooms in on the party politics orientations of the cabinet in office to understand how political priorities have been translated into R&I policy strategies. The second conceptual pillar focuses on the actors making policy strategies operational; hence, on those stakeholders identifying the substantive requirements of behaviors embedded in policy instruments' action. Design choices encompassing interdependent policy issues involve coordination across different policy sectors (Edler and Kuhlmann 2008; Griessen and Braun 2008; Koch 2008). Therefore, by illustrating the evolution of ministerial organizational layouts, the study investigates the way actors, in their interactions with institutional and organizational contexts, calibrate the characteristics of the inducement administered to policy recipients. The third conceptual pillar explains different typologies of interest intermediation strategies by describing the types of governing arrangements accompanying each instrument along its activation process. Policy design literature has mostly approached the study of this process from an "input perspective" (Howlett et al. 2020); hence, by looking at policy instrumentation simply as the result of decisions made by policymakers, while overlooking the role of policy recipients. However, policy instruments represent the building blocks of different governance agreements between policy takers and policymakers (Capano et al. 2019). Therefore, policy recipients do benefit from different policy instrument actions for undergoing their R&I activities, as much as policymakers benefit from their engagement in these activities. Consequently, the third pillar investigates how the organizational layouts of different interest intermediation strategies, in their interaction with the institutional setup, define different opportunity structures for recipients to influence policy design activities. Figure 2.4 illustrates this analytical perspective graphically.

Starting from the left side of Fig. 2.1, the political orientation of governments in power is used as a proxy for a more general understating of the political stakes related to alternative policy-mix scenarios. Indeed, the

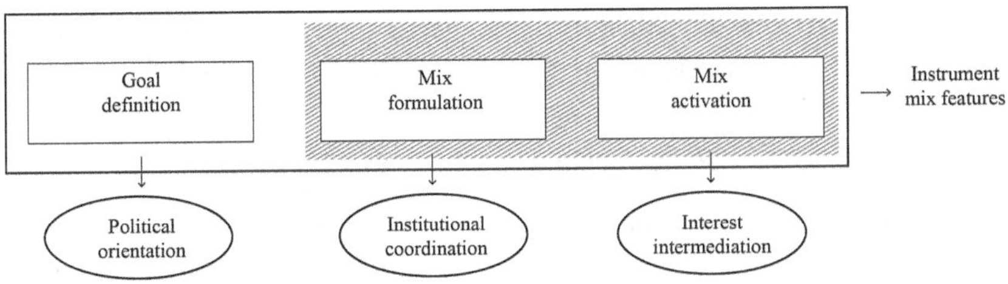

Fig. 2.4 Exploratory framework of analysis

political fractures activated by different interpretations of R&I problems and solutions shed light on the payoffs of alternative design choices. Moving to the right side of the table, the combination of actors' interactions—in terms of both internal (e.g., coordination between ministers/departments with R&I competences) and external (e.g., interaction with the target population) policymaking practices—describes the instrument calibration process. This clarifies how internal and external cooperation between policymakers and policy recipients influences the shape and the delivery structure, characterizing the final instrument mix.

This exploratory framework investigates the determinants of policy instrument selection according to the interactions between politics, interests, and institutions. It breaks up the analysis into a macro and micro level of analysis. The former investigates national partisan politics (political orientations of the cabinet in office), while the latter focuses on the interactions between ministerial organizational layouts (institutions) and interest intermediation strategies (interests). The choice to place the study of the goal definition at a higher level of abstraction is motivated by an investigation of the means-end rationalities of different cabinets. This allows us to reconstruct the architecture of choices from the perspective of decision-makers. Differently, institutions and interests stand at a lower level of abstraction, and they are analyzed together since they can jointly illustrate the way policymakers and policy takers interact—ultimately being "two sides of the same process."

This theoretical choice is also grounded in the literature on policy design (Howlett 2009) and policy changes (Hall 1993), which highlights the relevance of a theoretical differentiation between different levels, or orders, involved in policymaking activities. As further discussed in the methodology of this book (Chap. 3), this analytical strategy is also supported by the adoption of a different set of methodological procedures for data generation and analysis. In the following sections, the operationalization of these three theoretical pillars is discussed considering their contribution to a behavioral perspective on policy instruments' action.

2.3.1 *Political Orientations and Policy Types*

The possibility for new material conditions to trigger changes in policy design choices depends on the ability of current policy frames to provide solutions, or not, to new emerging problems (Sanz-Mendez and Borrás 2001). For this reason, design choices reflect political and social values of policymakers, as well as national trends and ideas about "good" policies (Schneider and Sidney 2009). As a consequence, the framings of policy problems, and their solutions, tend to cluster over time into long-term policy rationalities (Edler et al. 2016; Fagerberg 2017), which ultimately act as lenses filtering information and directing policymakers' attention toward different issues (Wilson 2000). These mechanisms lay the foundations for developing particular ways of designing policies, consolidating into specific policy types. However, while policymakers develop standard ways of designing policy, their impact still differs according to their purposes and consequences (Lowi 1964; Wilson 1973, 1980). Social groups relate to different public policy characteristics in indifferent ways (Freeman 2006), and this creates some contentiousness around their features and goals. Similarly, the way decision-makers identify problems and the way they want R&I to contribute to their solutions is also part of a political strategy (Edler and Fagerberg 2017). Therefore, policy types represent the results of societal demands requiring different costs and benefits distribution; and these can likely be associated with alternative patterns of politics. Public policies constitute the product of "competing societal demands that are prioritized and acted on by governments" (Choi et al. 2008:29). And having different parties in government indeed reflects the "transmission-belt between the state and the society" (Hall 1993:288).

The existing scholarship has demonstrated how the deployment of R&I instruments is able to alter the space of politics (Edquist and Borrás 2013). In addition, it shows how the political context can differently influence the availability of financial resources for R&D investments (Bergek et al. 2015) while still legitimizing different spaces in which policy recipients can gain access to resources (Van de Ven and Garud 1989). We know that R&I policy instruments are political devices because they display a high degree of interpretative flexibility (Flanagan et al. 2011). However, we still know little about the way alternative policy types (Ergas 1987) can explain different patterns of R&I governance trajectories. Policy instruments are chosen with a purpose (Edler et al. 2016); so, the way these are designed and implemented represents a manifestation of different political views. Hence, investigating the interaction between politics and the consolidated national R&I policy styles can explain the mechanisms shaping R&I policy goals' definition.

2.3.2 Institutional Coordination and Ministerial Organizational Layouts

Research and Innovation policies encompass different interdependent issues, which do not easily map onto the typical areas of activity of formal policymaking institutions (Pelkonen et al. 2008). This calls for the creation of organizations that can cope with issues at the intersection of "classic policy sectors." Consequently, the way policymaking competences have been organized reflects how national governments have differently framed R&I policy problems, their political priorities, and the policy strategy to address them (B G Peters 1998a, b; Tosun 2018).

Ministers often retain a central role in policy design within their areas of expertise (Laver et al. 1996); thus, sharing policymaking competences with peer organizations can ultimately alter the institutional power balance within governments (especially in the case of coalition governments). Similarly, more distributed policy competences do also require greater coordination between institutions to ensure the smooth coexistence between different policy strategies. Therefore, the way formal R&I competence divisions are enforced across ministries implies a scenario of interactions and dialogue between institutions or among different departments within the same institution (Braun 2008a). Such a distinction between the way policymaking institutions are officially organized and the way these are required to function lays down the theoretical foundations for what the literature defines as: formal versus informal institutions (Helmke and Levitsky 2004), formal versus informal administrative routines (Knill 2001), or administrative structures versus administrative styles (van Waarden 1995). Despite the different jargons, all these concepts point toward the same phenomenon characterizing the way state institutions work: on the one hand, institutions function following established norms and rules, while at the same time they also maintain a distinct character in the procedures by which they handle their daily tasks and duties (Bayerlein et al. 2021:153). Formal and informal institutions assist one another (Borrás 2009; Peters 2012) and represent the way institutions are not only structures adding stability to the political systems, but also interacting with the actors, interests, and ideas populating that system (Blyth 2003). However, we still know little about the way these two components interact in complex governance arrangements, like the case of R&I policies. Therefore, by investigating different ministerial organizational layouts across the policy design process, we can explore how the dual components of policymaking institutions support different scenarios of integration between policymaking responsibilities.

2.3.3 *Interest Intermediation Strategies and Actor Configurations*

Policy design is a social and material activity involving a heterogeneous constellation of actors (Voß and Simons 2018). This process defines governance environments where policymakers and policy takers become interdependent. The latter are defined as the individuals and groups whose behaviors are expected to be affected by policy activity (Howlett 2018:106), which is why this definition has been used interchangeably with policy targets or target population. Policy recipients do benefit from public support (in different forms) while undergoing their activities, as much as policymakers benefit from the results of these activities. Within this context, governing efforts to design effective policies face a collective action problem: policy takers should believe that the intervention implemented by governments is legitimate and that it is partially also in their self-interest to perform the expected behavior (Hofmann et al. 2014; Howlett 2018). Governments have increasingly lost their abilities to move unilaterally, neglecting preferences and resources of other social actors involved in policymaking (Bressers and O'Toole 2005; Fraussen 2014; Smits et al. 2010). Meanwhile, policy recipients have increasingly acquired the power to nullify, or alter, the operation of policy instruments (Woodside 1986). Indeed, actors' responsiveness (Trein 2017) can shape policy design because the lack of policy recipients' support, or legitimacy, for the behavioral change sought by policy instruments can hamper cooperation and undermine successful implementation (Curley et al. 2020; Dermont et al. 2017; Gross 2007; Ingold et al. 2018; Kammermann and Ingold 2019; Varone and Aebischer 2001).

Therefore, policy instruments represent the building blocks of different governance agreements between policy takers and policymakers (Capano et al. 2019) to the extent that policy tools define the division of competences between third parties and public actors in the enforcement of public action. Indeed, policy instruments organize governing strategies, shape practices, allocate roles, and create social positions in different ways, by distributing resources in different ways (Lascoumes and Le Gales 2007). However, target populations can involve a heterogeneous group of actors, with different attitudes, behaviors, and tools used, and especially with different expectations about the effect of policy-tool deployment (Schneider and Ingram 1990). This requires creating arrangements that "marry" private interest with public interests, or that can at minimum establish communication between policy takers and policymakers. Therefore, as complementary to institutional coordination, exploring the systems of interest intermediation between policymakers and policy takers can guide the investigation of the role of target groups in R&I policy design and the diachronic evolution of national policy mix.

2.3.4 Behavioral Changes and Policy Instruments: An Actor Perspective on Policy Takers and Policymakers

Exercising power means obtaining behaviors in line with our expectations (Stoppino 2001), and policy instruments substantiate the power to "get people do things they might not otherwise do" (Schneider and Ingram 1990:513). Therefore, it is critical to understand whether the behavioral change supported by a policy instrument is likely to trigger compliance or not. In the hypothetical scenario of automatic compliance, recipients would align with the suggested behavioral change and government activity would simply provide information about the expected behaviors (Howlett 2018). However, in reality, political institutions are experiencing increasing resource dependency, hence the inability to act unilaterally (Bressers and O'Toole 1998; Curley et al. 2020; Dermont et al. 2017; Goetz 2008; Gross 2007; Ingold et al. 2018; Kammermann and Ingold 2019; Varone and Aebischer 2001). Therefore, the most basic activities of governance also involve consideration on the part of the recipients regarding the appropriateness of government action (March and Olsen 2004), suggesting how instruments actually possess a value "shared" by different actors (Capano and Lippi 2017). Thus, understanding policy design choices requires us to investigate instrument choice from a two-way perspective. This can allow for an element of "supply push" in policy change (Voß and Simons 2018), where both policymakers and policy takers have the power to influence the likelihood of certain instruments' features to be selected.

For that purpose, this research adopts an operationalization of tool dimensions focusing on the drivers of expected behaviors induced by policy instruments, according to three hierarchical components. It focuses on the means-end rationalities embedded in instrument choices, on how social control is exercised (instrument shape), and on the types of governing arrangements coming together with the selection of different instruments (delivery structure). By introducing a theoretical and analytical differentiation between instrument families, shapes, and delivery components, we aim at overcoming the (restricted) focus on policy instruments as a means through which governments use their legitimate power to shape public action (instrument families). Indeed, any tool is actually a "package" of different elements that can be disaggregated into relatively small units, and scored according to different behavioral dimensions (Schneider and Ingram 1990). Establishing an instrument involves a number of different tasks, and each of them can attract the interest of different stakeholders. Thus, understanding policy design requires us to unravel the social dynamics emerging from these choices (Varone and Aebischer 2001; Voß and Simons 2018).

The innovative contribution of this study is namely to investigate policy design from the perspective of both the actors "making" and the actors "receiving" policy design choices. To understand the underlining forces influencing policy design choices, it is necessary to investigate this process from a two-way perspective, where both policymakers and policy takers have the power to influence the likelihood of certain instruments' features to be selected. As demonstrated in the literature (Capano et al. 2019), the mix of instrumental shapes created by specific policy design choices makes a difference regardless of the "quantity" of the shapes included. Indeed, despite the fact instruments may look similar in the way they deal with a problem, there will always be substantial differences in the concrete details of how they are chosen and designed in the context in which they are applied. Therefore, the dynamics embedded in the functioning of different instruments requires us to simultaneously account for different phenomena, like the dominance of a given policy style (2002, 2011), the political components embedded in the unfolding of instrument actions (Vedung 1998), as well as the power affected interests might have in this process (Woodside 1986).[8] This, combined with an analytical perspective focused on the behavioral changes required for the target population, can contribute to a deeper analytical understanding of policy design.

[8] These factors characterize the multidimensionality (Salamon 2002) embedded in policy instruments.

Bibliography

Acciai, Claudia. 2021. The Politics of Research and Innovation: Understanding Instrument Choices in Complex Governance Environments – the Case of France and Italy. *Research Policy* 50 (9): 104254.

Acciai, Claudia, and Giliberto Capano. 2021. Policy Instruments at Work: A Meta-Analysis of Their Applications. *Public Administration* 99: 118–136.

Alexander, G.L., and A. Bennett. 2005. *Case Studies and Theory Development.* Cambridge: MIT Press.

Arnold, Erik. 2007. CREST 3 % OMC Third Cycle Policy Mix Peer Review Country Report France.

Arnold, Erik, and Patries Boekholt. 2003. Research and Innovation Governance in Eight Countries: A Meta-Analysis of Work Funded by EZ (Netherlands) and RCN (Norway). Technolopolis.

Banfi, A., and G. Viesti. 2016. Il Finanziamento Delle Università. In *Università in declino. Un'indagine sugli atenei da Nord a Sud*, ed. G. Viesti, 277–304. Donzelli. Roma.

Baumgartner, Frank R., and Bryan D. Jones. 1993. *Agendas and Instability in American Politics.* Chicago: University of Chicago Press.

Bayerlein, Louisa, Christoph Knill, and Dionys Zink. 2021. The Concept of Administrative Styles. In *The Routledge Handbook of Policy Styles*, ed. M. Howlett and J. Tosun, 153–164. London: Routledge.

BCG & CM INTERNATIONAL. 2008. *Evaluation Des Pôles de Compétitivité, Synthèse Du Rapport d'évaluation.*

Béland, Daniel. 2009. Ideas, Institutions, and Policy Change. *Journal of European Public Policy* 16 (5): 701–718.

Béland, Daniel, and Michael Howlett. 2016. The Role and Impact of the Multiple-Streams Approach in Comparative Policy Analysis. *Journal of Comparative Policy Analysis: Research and Practice* 18 (3): 221–227.

Bergek, Anna, Marko Hekkert, Staffan Jacobsson, Jochen Markard, Björn Sandén, and Bernhard Truffer. 2015. Technological Innovation Systems in Contexts: Conceptualizing Contextual Structures and Interaction Dynamics. *Environmental Innovation and Societal Transitions* 16: 51–64.

Berger, Suzanne. 2016. Reforms in the French Industrial Ecosystem.

Bersani, Pier Luigi, Alfonso Lelio, and Letta Enrico. 2004. *Viaggio Nell'economia Italiana.* Donzelli.

Blyth, Mark. 2003. Structures Do Not Come with an Instruction Sheet: Interests, Ideas, and Progress in Political Science. *Perspectives on Politics* 1 (4): 695–706.

Bologna, Federica, Angelo Di Iorio, Silvio Peroni, and Francesco Poggi. 2021. *Do Open Citations Inform the Qualitative Peer-Review Evaluation in Research Assessments? An Analysis of the Italian National Scientific Qualification.*

Borrás, Susana. 2009. The Widening and Deepening of Innovation Policy: What Conditions Provide for Effective Governance? In *Papers in Innovation Studies 2009/2, Lund University, CIRCLE - Center for Innovation, Research and Competences in the Learning Economy.*

Borrás, Susana, and Jakob Edler. 2015. *The Governance of Socio-Technical Systems: Explaining Change.* Cheltenham: Edward Elgar Publishing.

Borrás, Susana, and Charles Edquist. 2019. *Holistic Innovation Policy.* Oxford University Press.

Borrás, Susana, and Leonard Seabrooke. 2015. Sources of National Institutional Competitiveness: Sensemaking in Institutional Change.

Bouwma, Irene, Bas Arts, and Duncan Liefferink. 2017. Cause, Catalyst or Conjunction? The Influence of the Habitats Directive on Policy Instrument Choice in Member States. *Journal of Environmental Planning and Management* 60 (6): 977–996.

Boyatzis, Richard E. 1998. *Transforming Qualitative Information: Thematic Analysis and Code Development.* Thousand Oaks: Sage Publications.

Braun, Dietmar. 1993. Who Governs Intermediary Agencies? Principal-Agent Relations in Research Policy-Making. *Journal of Public Policy* 13 (02): 135.

———. 2008a. Lessons on the Political Coordination of Knowledge and Innovation Policies. *Science and Public Policy* 35 (4): 289–298.

———. 2008b. Organising the Political Coordination of Knowledge and Innovation Policies. *Science and Public Policy* 35 (4): 227–239.

Bressers, H.A., and L. O'Toole. 2005. Instrument Selection and Implementation in a Networked Context. In *Designing Government: From Instruments to Governance*, ed. P. Eliadis, M. Hill, and M. Howlett, 132–153. Montreal: M.-Q. U. Press.

Bressers, Hans, and Laurence O'Toole. 1998. The Selection of Policy Instruments: A Network-Based Perspective. *Journal of Public Policy* 18: 213–239.

Buchs, Milena. 2007. *New Governance in European Social Policy: The Open Method of Coordination.* Palgrave Macmillan.

Burns, J.M.C. 2010. Cross-Case Synthesis and Analysis. In *Encyclopedia of Case Study Research*, ed. A.J. Mills, G. Durepos, and E. Wiebe, vol. 1. Thousand Oaks: SAGE Publications.

Cantner, Uwe, and Andreas Pyka. 2001. Classifying Technology Policy from an Evolutionary Perspective. *Research Policy* 30 (5): 759–775.

Capano, G., J.E. Rayner, and A.R. Zito. 2012. Governance from the Bottom up: Complexity and Divergence in Comparative Perspective. *Public Administration* 90 (1): 56–73.

Capano, Giliberto., and Elisabetta, Gualmini. 2011. Le Pubbliche Amministrazioni in Italia.

Capano, Giliberto. 2008. Looking for Serendipity: The Problematical Reform of Government within Italy's Universities. *Higher Education* 55 (4): 481–504.

Capano, Giliberto, and Andrea Lippi. 2017. How Policy Instruments Are Chosen: Patterns of Decision Makers' Choices. *Policy Sciences* 50 (2): 269–293.

Capano, Giliberto, Andrea Pritoni, and Giulia Vicentini. 2019. Do Policy Instruments Matter? Governments' Choice of Policy Mix and Higher Education Performance in Western Europe. *Journal of Public Policy* 40(3): 375–401.

Capano, Giliberto, Marino Regini, and Matteo Turri. 2016. *Changing Governance in Universities: Italian Higher Education in Comparative Perspective*. London: Palgrave Macmillan.

Cashore, Benjamin, and Michael Howlett. 2007. Punctuating Which Equilibrium? Understanding Thermostatic Policy Dynamics in Pacific Northwest Forestry. *American Journal of Political Science* 51 (3): 532–551.

Chandler, J.A. 2014. France. In *Comparative Public administration*. London: Routledge.

Chiang, Jong-Tsong. 1991. From 'Mission-Oriented' to 'Diffusion-Oriented' Paradigm: The New Trend of U.S. Industrial Technology Policy. *Technovation* 11 (6): 339–356.

Chmiliar, L. 2010. "Multiple Case Design." in Mills, A. J., Durepos, G., Wiebe, E. (Eds.). Encyclopedia *of Case Study Research*. Thousand Oaks: SAGE Publications.

Choi, Sang Ok, Sang-Seok Bae, Sung-Wook Kwon, and Richard Feiock. 2008. County Limits: Policy Types and Expenditure Priorities. *The American Review of Public Administration* 40 (1): 29–45.

Chou, Meng-Hsuan, Jens Jungblut, Pauline Ravinet, and Martina Vukasovic. 2017. Higher Education Governance and Policy: An Introduction to Multi-Issue, Multi-Level and Multi-Actor Dynamics. *Policy and Society* 36 (1): 1–15.

Chung, Chao-chen. 2013. Government, Policy-Making and the Development of Innovation System: The Cases of Taiwanese Pharmaceutical Biotechnology Policies (2000–2008). *Research Policy* 42 (5): 1053–1071.

Cobis, Fabrizio. 2008. I Nuovi Processi Di Innovazione e l'azione Del Ministero Dell'Università e Della Ricerca. In *I Nuovi Processi Di Innovazione e l'azione Del Ministero Dell'Università e Della Ricerca*, 1000–1005.

Coccia, Mario, and Secondo Rolfo. 2010. New Entrepreneurial Behaviour of Public Research Organisations: Opportunities and Threats of Technological Services Supply. *International Journal of Services Technology and Management* 13 (1/2): 134–151.

Considine, Mark, and Kamran Ali Afzal. 2010. Legitimacy. In *The SAGE Handbook of Governance*, ed. M. Bevir. Berkeley: University of California, SAGE Publications.

Corradi, Fiammetta. 2009. *Alla Ricerca Dell'eccellenza: Le Politiche per l'eccellenza Nell'istruzione Superiore in Quattro Paesi Europei*. LED, Edizioni universitarie di lettere, economia, diritto.

Curley, Cali, Richard Feiock, and Xu. Kewei. 2020. Policy Analysis of Instrument Design: How Policy Design Affects Policy Constituency. *Journal of Comparative Policy Analysis: Research and Practice* 22 (6): 536–557.

Cytermann, Jean-Richard. 2006. L'architecture de la loi organique relative aux lois de finances (LOLF) dans les domaines de l'éducation et de la recherche: choix politiques ou choix techniques ? *Revue française d'administration publique* 117 (1): 85–93.

Dente, B. 2014. *Understanding Policy Decisions*. Cham: Springer.

Dermont, Clau, Karin Ingold, Lorenz Kammermann, and Isabelle Stadelmann-Steffen. 2017. Bringing the Policy Making Perspective in: A Political Science Approach to Social Acceptance. *Energy Policy* 108: 359–368.

Doern, G., and R. Phidd. 1983. *Canadian Public Policy: Ideas, Structures, Process*. Toronto: Methuen.

Donina, Davide, Michele Meoli, and Stefano Paleari. 2015. Higher Education Reform in Italy: Tightening Regulation Instead of Steering at a Distance. *Higher Education Policy* 28 (2): 215–234.

Dosso, Mafini. 2014. "Restructuring in France's Innovation System: From the Mission-Oriented Model to a Systemic Approach of Innovation." *LEM Papers Series*.

Dür, Andreas. 2008. Measuring Interest Group Influence in the EU: A Note on Methodology. *European Union Politics* 9 (4): 559–576.

Edler, Jakob, Paul Cunningham, Kieron Flanagan, and Philippe Laredo. 2013. *Innovation Policy Mix and Instrument Interaction: A Review*. London: Nesta.

Edler, Jakob, Paul Cunningham, A. Gök, and Philip Shapira. 2016. *Handbook of Innovation Policy Impact*. Cheltenham: Edward Elgar Publishing.

Edler, Jakob, and Jan Fagerberg. 2017. Innovation Policy: What, Why, and How. *Oxford Review of Economic Policy* 33 (1): 2–23.

Edler, Jakob, and Stefan Kuhlmann. 2008. Coordination within Fragmentation: Governance in Knowledge Policy in the German Federal System. *Science and Public Policy* 35 (4): 265–276.

Edquist, Charles. 2001. The Systems of Innovation Approach and Innovation Policy: An Account of the State of the Art. In DRUID conference, Aalborg (pp. 12–15).

Edquist, Charles, and Susana Borrás. 2013. The Choice of Innovation Policy Instruments. *Technological Forecasting and Social Change* 80: 1513–1522.

Eparvier, Patrick; Turcat, Nicolas; Schoen, Antonie; Carat, Gerard; Nill, Jan. 2008. ERAWATCH Country Report 2008. An Assessment of Research System and Policies France

Eparvier, Patrick, Flora Giarracca, and Léonor Rivoire. 2009. *ERAWATCH Country Report 2009: France*.

Eparvier, Patrick, Olivier Mallet, and Léonor Rivoire. 2011. *ERAWATCH Country Reports 2011: France*.

Ergas, H. 1987. The Importance of Technology Policy. In *Economic Policy and Technological Performance*, ed. P. Dasgupta and P. Stoneman. Cambridge: Cambridge University Press.

Etzioni A 1961. *Comparative Analysis of Complex Organizations.* The Free Press, New York.

European Commission. 1999. European Trend Chart on Innovation Country Report: Italy.

———. 2000a. European Trend Chart on Innovation - Country Report France.

———. 2000b. European Trend Chart on Innovation Country Report: Italy.

———. 2001. European Trend Chart on Innovation Country Report: Italy.

———. 2002a. European Trend Chart on Innovation Country Report Italy.

———. 2002b. EUROPEAN TREND CHART Theme-Specific Country Report: France.

———. 2003. European Trend Chart on Innovation Country Report: Italy.

———. 2004a. European Trend Chart on Innovation Annual Innovation Policy Trends and Appraisal Report: France.

———. 2004b. European Trend Chart on Innovation Country Report: Italy.

———. 2005a. European Trend Chart on Innovation - Annual Innovation Policy Trends and Appraisal Report: France.

———. 2005b. European Trend Chart on Innovation Country Report: Italy.

———. 2006. European Trend Chart on Innovation Country Report: Italy.

Evangelista, Rinaldo. 2007. Rilevanza e Impatto Delle Politiche Dell'innovazione in Italia. I Risultati Delle Indagini CIS. *Economia e Politica Industriale* 1: 103–124.

Fagerberg, Jan. 2017. Innovation Policy: Rationales, Lessons and Challenges. *Journal of Economic Surveys* 31 (2): 497–512.

Flanagan, Kieron, Elvira Uyarra, and Manuel Laranja. 2011. Reconceptualising the 'Policy Mix' for Innovation. *Research Policy* 40 (5): 702–713.

Foray, D., D.C. Mowery, and R.R. Nelson. 2012. Public R&D and Social Challenges: What Lessons from Mission R&D Programs? *Research Policy* 41 (10): 1697–1702.

Foray, Dominique. 2009. *The New Economics of Technology Policy*. Cheltenham: Edward Elgar. London.

Fraussen, Bert. 2014. The Visible Hand of the State: On the Organizational Development of Interest Groups. *Public Administration* 92 (2): 406–421.

Freeman, Gary P. 2006. National Models, Policy Types, and the Politics of Immigration in Liberal Democracies. *West European Politics* 29 (2): 227–247.

French Court of Auditors. 2013. Le Financement Public de La Recherche, Un Enjeu National.

Gallo, Riccardo. 2016. Torniamo a Industriarci: A Novant'anni Dalla 'Grande Crisi' Guida.

Gallo, Riccardo, Mauro Mallone, and Vincenzo Zezza. 2008. Technology Transfer: The RIDITT Experience. *L'industria* 1: 207–226.

Gallo, Riccardo, and Francesco Silva. 2006. Le Condizioni per Crescere: Diagnosi e Proposte per Il Sistema Produttivo. Il Sole 24 ore.

Gassler, Helmut, Wolfgang Polt, and Christian Rammer. 2008. Priority Setting in Technology Policy: Historical Developments and Recent Trends. In *Innovation Policy in Europe: Measurement and Strategy*, ed. C. Nauwelaers and R. Wintjes. Cheltenham: Edward Elgar.

Giest, Sarah. 2016. The Challenges of Enhancing Collaboration in Life Science Clusters: Lessons from Chicago, Copenhagen and Singapore: Table 1. *Science and Public Policy* 44 (2): scw046.

Goetz, Klaus H. 2008. Governance as a Path to Government. *West European Politics* 31 (1–2): 258–279.

Griessen, Thomas, and Dietmar Braun. 2008. The Political Coordination of Knowledge and Innovation Policies in Switzerland. *Science and Public Policy* 35 (May): 277–288.

Gross, Catherine. 2007. Community Perspectives of Wind Energy in Australia: The Application of a Justice and Community Fairness Framework to Increase Social Acceptance. *Energy Policy* 35 (5): 2727–2736.

Hall, Peter A. 1993. Policy Paradigms, Social Learning, and the State: The Case of Economic Policymaking in Britain. *Comparative Politics* 25 (3): 275–296.

———. 2010. Historical Institutionalism in Rationalist and Sociological Perspective. In *Explaining Institutional Change: Ambiguity, Agency, and Power*, ed. J. Mahoney and K. Thelen, 204–224. Cambridge: Cambridge University Press.

Hartmann, Eva, and Poul F. Kjaer. 2015. *The Evolution of Intermediary Institutions in Europe*. London: Palgrave Macmillan UK.

Helmke, Gretchen, and Steven Levitsky. 2004. Informal Institutions and Comparative Politics: A Research Agenda. *Perspectives on Politics* 2 (4): 725–740.

Héritier, Adrienne, and Dirk Lehmkuhl. 2011. New Modes of Governance and Democratic Accountability. *Government and Opposition* 46 (1): 126–144.

Hofmann, Eva, Katharina Gangl, Erich Kirchler, and Jennifer Stark. 2014. Enhancing Tax Compliance through Coercive and Legitimate Power of Tax Authorities by Concurrently Diminishing or Facilitating Trust in Tax Authorities. *Law and Policy* 36 (3): 290–313.

Hood, Christopher. 1986. *The Tools of Government / Christopher C. Hood*. Chatham: Chatham House.

Hood, Christopher, and Helen Magretts. 2007. *The Tools of Government in the Digital Age*. Macmillan. Red Globe Press.

Howlett, M., and Ramesh, M. 2003. *Studying Public Policy: Policy Cycles and Policy Subsystems*. Oxford University Press.

Howlett, M., and I. Mukherjee. 2018. *Routledge Handbook of Policy Design*. 1st ed. London: Routledge.

Howlett, M., M. Ramesh, and A. Perl. 2009. *Studying Public Policy: Policy Cycles & Policy Subsystems*. Oxford: Oxford University Press.

Howlett, Michael. 2000. Managing the 'Hollow State': Procedural Policy Instruments and Modern Governance. *Canadian Public Administration/Administration Publique Du Canada* 43 (4): 412–431.

———. 2009. Governance Modes, Policy Regimes and Operational Plans: A Multi-Level Nested Model of Policy Instrument Choice and Policy Design. *Policy Sciences* 42 (1): 73–89.

———. 2018. Aligning Policy Tools and Their Targets: Nudging and Utility Maximization in Policy Design. In *Routledge Handbook of Policy Design*, ed. M. Howlett and I. Mukherjee, 1st ed., 106–121. *London*: Routledge.

Howlett, Michael, and Ishani Mukherjee. 2014. Policy Design and Non-Design: Towards a Spectrum of Policy Formulation Types. *Politics and Governance* 2 (2): 57.

Howlett, Michael, and M. Ramesh. 1993. Patterns of Policy Instrument Choice: Policy Styles, Policy Learning and the Privatization Experience. *Review of Policy Research* 12 (1–2): 3–24.

Howlett, Michael, M. Ramesh, and Giliberto Capano. 2020. Policy-Makers, Policy-Takers and Policy Tools: Dealing with Behaviourial Issues in Policy Design. *Journal of Comparative Policy Analysis: Research and Practice* 22 (6): 487–497.

Howlett, Michael, and Jeremy Rayner. 2013. Patching vs Packaging in Policy Formulation: Assessing Policy Portfolio Design. *Politics and Governance* 1 (2): 170.

IGAENR. 2012. *Etude Des Mécanismes d'allocation Des Moyens Humains et Financiers Aux Unités de Recherche Par Les Organismes de Recherche*. Direction de l'information légale et administrative.

Ingold, Karin, Isabelle Stadelmann-Steffen, and Lorenz Kammermann. 2018. The Acceptance of Instruments in Instrument Mix Situations: Citizens' Perspective on Swiss Energy Transition. *Research Policy* 48: 103694.

John, Peter. 2012. *Analysing Public Policy*. 2nd ed. London: Routledge.

Jungblut, Jens. 2015. Bringing Political Parties into the Picture: A Two-Dimensional Analytical Framework for Higher Education Policy. *Higher Education* 69 (5): 867–882.

Kammermann, Lorenz, and Karin Ingold. 2019. Going beyond Technocratic and Democratic Principles: Stakeholder Acceptance of Instruments in Swiss Energy Policy. *Policy Sciences* 52 (1): 43–65.

Kassim, Hussein, and Patrick Le Galès. 2010. Exploring Governance in a Multi-Level Polity: A Policy Instruments Approach. *West European Politics* 33 (1): 1–21.

Knill, Christoph. 2001. *The Europeanisation of National Administrations: Patterns of Institutional Change and Persistence*. Cambridge: Cambridge University Press.
Koch, Christian. 2008. The Superministry Approach: Integrated Governance of Science, Technology and Innovation with Contracted Autonomy. *Science and Public Policy* 35 (4): 253–264.
Lai, Luigi. 2016. Innovation Policies and SMEs Development. In *Entrepreneurship: Antecedents and Effects "Przedsiębiorczość Międzynarodowa"*, ed. M. Kosała, M. Urbaniec, and A. Żur, vol. 2, no. 2, 147–163. Kraków: Cracow University of Economics.
Lanza, Orazio, and Kostas Lavdas. 2000. The Disentanglement of Interest Politics: Business Associability, the Parties and Policy in Italy and Greece. *European Journal of Political Research* 37 (2): 203–235.
Laranja, Manuel, Elvira Uyarra, and Kieron Flanagan. 2008. Policies for Science, Technology and Innovation: Translating Rationales into Regional Policies in a Multi-Level Setting. *Research Policy* 37 (5): 823–835.
Larrue, Philippe, Patrick Eparvier, and Sophie Bussillet. 2006. *Etude de l'impact Du Crédit Impôt Recherche*.
Lascoumes, Pierre, and Patrick Le Gales. 2007. Introduction: Understanding Public Policy through Its Instruments? From the Nature of Instruments to the Sociology of Public Policy Instrumentation. *Governance* 20 (1): 1–21.
Laver, Michael, and Kenneth Shepsle. 1996. The Social Context of Government Formation. In *Making and Breaking Governments. Cabinets and legislatures in Parliamentary Democracies*, ed. K. Laver and Shepsle Michael, 18–42. Cambridge: Cambridge University Press.
Lepori, Benedetto. 2011. Coordination Modes in Public Funding Systems. *Research Policy* 40 (3): 355–367.
Lepori, Benedetto, Emanuela Reale, and Emilia Primieri. 2017. *Public Funding Country Profile France Annex 11*.
Lindblom, C.E. 1959. The Science of 'Muddling Through.'. *Public Administration Review* 19 (2): 79–88.
Linder, Stephen H., and B. Guy Peters. 1989. Instruments of Government: Perceptions and Contexts. *Journal of Public Policy* 9 (01): 35.
Lissoni, Francesco, Peter Lotz, Jens Schovsbo, and Adele Treccani. 2009. Academic Patenting and the Professor's Privilege: Evidence on Denmark from the KEINS Database. *Science and Public Policy* 36 (8): 595–607.
Lissoni, Francesco, and Fabio Montobbio. 2015. The Ownership of Academic Patents and Their Impact. Evidence from Five European Countries. *Revue économique* 66 (1): 143–171.
Lowi, T.J. 1964. American Business, Public Policy, Case Studies, and Political Theory. *Word Politics* 16: 677–715.

Lucchese, Matteo, Leopoldo Nascia, Mario Pianta, and Mario Pianta. 2016. Industrial Policy and Technology in Italy. *Economia e Politica Industriale* 43 (3): 233–260.

Macdonald, Douglas. 2001. Coerciveness and the Selection of Environmental Policy Instruments. *Canadian Public Administration/Administration Publique Du Canada* 44 (2): 161–187.

De Maio, Adriano. 2011. *L' Innovazione Vincente*. Brioschi.

Di Maio, Michele. 2014. The Italian Economy, the Economic Crisis and Industrial Policy. In *Structural change, competitiveness and industrial policy: painful lessons from the European periphery*, ed. A. Teixeira and E. Silva. London: Routledge.

Majone, Giandomenico. 1989. *Evidence, Argument, and Persuasion in the Policy Process*. New Heaven: Yale University Press.

———. 1976. Choice among Policy Instruments for Pollution Control. *Policy Analysis* 2: 589–613.

Malerba, Franco. 1993. The National System of Innovation: Italy. In *National Innovation Systems: A Comparative Analysis*.

March, James G., and Johan P. Olsen. 1997. *Democratic Governance*. New York: Free Press.

———. 2004. *The Logic of Appropriateness*. Oslo: Arena.

Mathison, Sandra. 2005. Cross-Case Analysis. In *Encyclopedia or Evaluation*, ed. S. Mathison. Thousand Oaks: SAGE Publications.

Merton, R.K. 1938. Social Structure and Anomie. *American Sociological Review* 3 (5): 672–682.

Miles, M.B., and A.M. Huberman. 1994. *Qualitative Data Analysis: An Expanded Sourcebook*. Thousand Oaks: Sage Publications, Inc.

Ministére de l'education nationale de l'enseignment supérieur et de la Recherche. 2014. *Développement et Impact Du Crédit d'impôt Recherche: 1983-2011*.

Ministero dell'Istruzione dell'Università e della Ricerca. 2005. *Programma Nazionale per La Ricerca*.

MISE. 2010. *Gli Incentivi Alle Attività Produttive. Effetti Della Legge 46/82 Sull'innovazione e Sull'industrializzazione Dei Prototipi Di Ricerca*.

Musselin, Christine (1958–........). 2017. *La Grande Course Des Universités*. edited by Sciences Po Presse.

Musselin, Christine, and Catherine Paradeise. 2009. France: From Incremental Transitions to Institutional Change. In *University governance: Western European comparative perspectives*, ed. Catherine Paradeise, Emanuela Reale, Ivar Bleiklie, and Ewan Ferlie, 21–49. Cham: Springer Verlag.

Mustar, Philippe, and Philippe Laredo. 2002. Innovation and Research Policy in France (1980-2000) or the Disappearance of the Colbertist State. *Research Policy* 31 (1): 55–72.

Mustar, Philippe, and Philippe Larédo. 2002. Innovation and Research Policy in France (1980–2000) or the Disappearance of the Colbertist State. *Research Policy* 31 (1): 55–72.
Nascia, Leopoldo, and Mario Pianta. 2013. *ERAWATCH Country Reports 2013: Italy.*
Nascia, Leopoldo, Bianca Potì, and Emanuela Reale. 2012. *ERAWATCH Country Report 2012: Italy.*
Nauwelaers, Claire, and Wintjes René. 2008. *Innovation Policy in Europe: Measurement and Strategy.* Cheltenham: Edward Elgar.
Nelson, Richard R. 1993. *National Innovation Systems: A Comparative Analysis.* New York: Oxford University Press.
North, D.C. 1990. *Institutions, Institutional Change and Economic Performance.* Cambridge: Cambridge University Press.
OECD. 2005. *Governance of Innovation Systems.* OECD.
———. 2014. *Reviews of Innovation Policy: France 2014.* OECD.
Olsson, Jan, and Joachim Åström. 2003. Why Regionalism in Sweden? *Regional & Federal Studies* 13 (3): 66–89.
Palier, Bruno. 2007. Tracking the Evolution of a Single Instrument Can Reveal Profound Changes: The Case of Funded Pensions in France. *Governance* 20 (1): 85–107.
Pelkonen, Antti, Tuula Teräväinen, and Suvi-Tuuli Waltari. 2008. Assessing Policy Coordination Capacity: Higher Education, Science and Technology Policies in Finland. *Science and Public Policy* 35 (4): 241–252.
Perry, Beth, and Tim May. 2007. Governance, Science Policy and Regions: An Introduction. *Regional Studies* 41 (8): 1039–1050.
Peters, B.G. 1998a. *Managing Horizontal Government: The Politics of Coordination.* Canadian Centre for Management Development.
Peters, B. Guy. 1998b. *Comparative Politics.* London: Macmillan Education UK.
———. 2000. The Politics of Tool Choice. In *The Tools of Government: A Guide to the New Governance*, ed. L. Salamon, 1st ed. New York: Oxford University Press.
———. 2012. Governance as Political Theory. In *The Oxford Handbook of Governance*, ed. Levi-Faur. New York: Oxford University Press.
Pierantozzi, Valentina. 2007. *ERAWATCH Analytical Country Report 2007: Italy.*
Pochet, Philippe, and Caroline de la Porte. 2002. *Building Social Europe through the Open Method of Co-Ordination.* Brussels: P. Lang.
Potì, Bianca, and Emanuela Reale. 2006. *Project Funding. Italy.*
———. 2007. Changing Allocation Models for Public Research Funding: An Empirical Exploration Based on Project Funding Data. *Science and Public Policy* 34 (6): 417–430.

———. 2010. *ERAWATCH Country Reports 2010: Italy.*

———. 2011. *ERAWATCH Country Report 2011: Italy.*

Poti, Bianca, Emanuela Reale, and Valentina Pierantozzi. 2008. *ERAWATCH Country Report. An Assessment of Research System and Policies. Italy.*

Prange, Heiko. 2008. Explaining Varieties of Regional Innovation Policies in Europe. *European Urban and Regional Studies* 15 (1): 39–52.

Reale, Emanuela, and Lucio Morettini. 2017. *Public Funding Country Profile Italy Annex 11.*

Reale, Emanuela, and Bianca Potì. 2009. Italy: Local Policy Legacy and Moving to an 'In Between' Configuration. In *University governance: Western European comparative perspectives*, ed. Catherine Paradeise, Emanuela Reale, Ivar Bleiklie, and Ewan Ferlie, 77–102. Cham: Springer Verlag.

Reillon, V. 2015. *Overview of EU Funds for Research and Innovation.* EPRS: European Parliamentary Research Service. Belgium.

Rimkutė, Dovilė, and Markus Haverland. 2015. How Does the European Commission Use Scientific Expertise? Results from a Survey of Scientific Members of the Commission's Expert Committees. *Comparative European Politics* 13 (4): 430–449.

Ringeling, Arthur B. 2005. Instruments in Four: The Elements of Policy Design. In *Designing Government: From Instruments to Governance*, ed. P. Eliadis, M.M. Hills, and M. Howlett, 185–202. Kingston: O. McGill-Queens University Press.

Rist, Ray C. 1998. Choosing the Right Policy Instrument at the Right Time: The Contextual Challenges of Selection and Implementation. In *In Carrots, Sticks and Sermons: Policy Instruments and Their Evaluation*, ed. M.-L. Bemelmans-Videc, R. Rist, and E. Vedung. New Brunswick: Transaction.

Ritchie, J., and J. Lewis. 2003. *Qualitative Research Practice: A Guide for Social Science Students and Researchers.* Thousand Oaks: SAGE Publications.

Rogge, Karoline S. 2018. Designing Complex Policy Mixes. In *Routledge Handbook of Policy Design*, ed. M. Howlett and I. Mukherjee, 1st ed., 34–58. London: Routledge.

Rogge, Karoline S., and Kristin Reichardt. 2016. Policy Mixes for Sustainability Transitions: An Extended Concept and Framework for Analysis. *Research Policy* 45 (8): 1620–1635.

Rueschemeyer, Dietrich, and John D. Stephens. 1997. Comparing Historical Sequences: A Powerful Tool for Causal Analysis. *Comparative Social Research* 16: 55–72.

Sabatier, Paul A. 1986. Top-Down and Bottom-Up Approaches to Implementation Research: A Critical Analysis and Suggested Synthesis. *Journal of Public Policy* 6 (1): 21–48.

Salamon, Lester. 2002. *The Tools of Government: A Guide to the New Governance.* Oxford University Press.

———. 2011. The New Governance and the Tools of Public Action. 28(5). Symposium - Redefining the Public Sector: Accountability and Democracy in the Era of Privatization.

Sanz-Mendez, Luis, and Susana Borrás. 2001. Explaining Changes and Continuity in EU Technology Policy: The Politics of Ideas. In *Changing European Research System*, ed. Imon Dresner and Nigel Gilbert. Aldershot: Ashgate Press.

Schmidt, Tobias S., and Sebastian Sewerin. 2019. Measuring the Temporal Dynamics of Policy Mixes – An Empirical Analysis of Renewable Energy Policy Mixes' Balance and Design Features in Nine Countries. *Research Policy* 48 (10): 103557.

Schneider, Anne, and Helen Ingram. 1990. Behavioral Assumptions of Policy Tools. *The Journal of Politics* 52 (2): 510–529.

———. 1993. Social Construction of Target Populations: Implications for Politics and Policy. *American Political Science Review* 87 (2): 334–347.

Schneider, Anne, and Mara Sidney. 2009. What Is Next for Policy Design and Social Construction Theory? *Policy Studies Journal* 37 (1): 103–119.

Schoen, Antonie, Gerard Carat, and Nill Jan. 2008. *ERAWATCH Analytical Country Report 2007: France.*

Schreier, Margrit. 2012. *Qualitative Content Analysis in Practice.* London: SAGE Publications Ltd.

Seawright, Jason, and John Gerring. 2008. Case Selection Techniques in Case Study Research: A Menu of Qualitative and Quantitative Options. *Political Research Quarterly* 61 (2): 294–308.

Seravalli, G. 2009. *Competitive European Regions through Research and Innovation. Different Theoretical Approaches to Innovation Policies.*

Silva, Francesco. 2007. La Politica Industriale Del Governo Prodi. *Economia e Politica Industriale* 2: 7–19.

Simons, Arno, and Jan-Peter Voß. 2018. The Concept of Instrument Constituencies: Accounting for Dynamics and Practices of Knowing Governance. *Policy and Society* 37 (1): 14–35.

Sirilli, Giorgio. 2010. *La Produzione e La Diffusione Della Conoscenza. Ricerca, Innovazione e Risorse Umane.*

Smits, Ruud, Philip Shapira, and Stefan Kuhlmann. 2010. *The Theory and Practice of Innovation Policy: An International Research Handbook.* Cheltenham: Edward Elgar.

Spence, R.E. 2014. Italy. In *Comparative public administration*, ed. J.A. Chandler, 2nd ed. London: Routledge, Taylor & Francis Group.

Stoppino, Mario. 2001. *Potere e Teoria Politica.* Giuffre.

Tamtik, Merli. 2016. Institutional Change Through Policy Learning: The Case of the European Commission and Research Policy. *Review of Policy Research* 33 (1): 5–21.

Thèves, Jean, Benedetto Lepori, and Philippe Larédo. 2007. Changing Patterns of Public Research Funding in France. *Science and Public Policy* 34 (6): 389–399.

Tödtling, Franz, and Michaela Trippl. 2005. One Size Fits All? Towards a Differentiated Regional Innovation Policy Approach. *Research Policy* 34: 1203–1219.

———. 2012. Transformation of Regional Innovation Systems: From Old Legacies to New Development Paths.

Tosun, Jale. 2018. Investigating Ministry Names for Comparative Policy Analysis: Lessons from Energy Governance. *Journal of Comparative Policy Analysis: Research and Practice* 20 (3): 324–335.

Traù, Fabrizio. 2009. Logica Di Un Sistema Di Sostegno Alle Imprese. La Politica Industriale Italiana Dal Secondo Dopoguerra a 'Industria 2015'. *Economia e Politica Industriale* 4.

Trebilcock, Michael J., and Douglas G. Hartle. 1982. The Choice of Governing Instrument. *International Review of Law and Economics* 2 (1): 29–46.

Trein, Philipp. 2017. A New Way to Compare Horizontal Connections of Policy Sectors: 'Coupling' of Actors, Institutions and Policies. *Journal of Comparative Policy Analysis: Research and Practice* 19 (5): 419–434.

Van de Ven, A.H., and R. Garud. 1989. A Framework for Understanding the Emergence of New Industries. In *Research on Technological Innovation and Management Policy*, ed. R. Rosenbloom and R. Burgelman. Greenwich: JAI Press.

van der Doelen, F., 1998. The 'give-and-take' packaging of policy instruments: optimising legitimacy and effectiveness. In *Carrots, Sticks and Sermons: Policy Instruments and Their Evaluation*, eds. M-L. Bemelmans-Videc, R. Rist, and E. Vedung. Transaction: New Brunswick.

Van der Meulen, Barend. 1998. Science Policies as Principal–Agent Games: Institutionalization and Path Dependency in the Relation between Government and Science. *Research Policy* 27 (4): 397–414.

Varone, Frédéric, and Bernard Aebischer. 2001. Energy Efficiency: The Challenges of Policy Design. *Energy Policy* 29 (8): 615–629.

Vedung, Evert. 1998. Policy Instruments: Typologies and Theories. In *Carrots, Sticks and Sermons: Policy Instruments and Their Evaluation*, ed. E. Vedung, M.L. Bemelmans-Videc, and R.C. Rist, 21–58. New Brunswick: Transaction Publishers.

Verhoest, Koen, and Geert Bouckaert. 2005. Machinery of Government and Policy Capacity. In *Challenges to State Policy Capacity*, ed. M. Painter and J. Pierre. Basingstoke: Palgrave.

Voß, Jan-Peter, and Arno Simons. 2014. Instrument Constituencies and the Supply Side of Policy Innovation: The Social Life of Emissions Trading. *Environmental Politics* 23 (5): 735–754.

———. 2018. Instrument Constituencies: Promoting Policy Designs. In *Routledge Handbook of Policy Design*, ed. M. Howlett and I. Mukherjee, 1st ed., 180–200. London: Routledge.

van Waarden, F. 1995. Persistence of National Policy Styles: A Study of Their Institutional Foundations. In *Convergence or Diversity*, ed. B. Unger and F. van Waarden, 333–372. Vermont: Ashgate Publishing Company.

Wilson, Carter. 2000. Policy Regimes and Policy Change. *Journal of Public Policy* 20 (3): 247–274.

Wilson, James. 1973. *Political Organizations*. New York: Basic Book.

———. 1980. *The Politics of Regulation*. New York: Basic Book.

Woodside, Kenneth. 1983. The Political Economy of Policy Instruments: Tax Expenditures and Subsidies. In *The Politics of Canadian Public Policy*, ed. Michael M. Atkinson and Marsha A. Chandler, 173–197. Toronto: University of Toronto Press.

———. 1986. Policy Instruments and the Study of Public Policy. *Canadian Journal of Political Science / Revue Canadienne de Science Politique* 19 (4): 775–793.

Zaparucha, Elisabeth. 2010. *ERAWATCH Country Report 2010: France*.

CHAPTER 3

Choosing and Blending Instruments for R&I Policies

The design of Research & Innovation (R&I) policy, the selection of instruments, and their blending into mixes represent an inherently political process. Decisions are taken within an environment where: "choices are shaped by the characteristics of the instruments, the nature of the problem, past governing experiences, subjective preferences of policymakers and the likely reaction of affected social groups" (Howlett and Ramesh 1993:13). Hence, conflicts and compromises between actors, the rule system inherited from the past, as well as the formulation of different cognitive and normative framings influence the output of this process (Linder and Guy Peters 1989; Salamon 2002, 2011). This chapter focuses on discussing the combined influence of these factors for policy design across two case studies: France and Italy. This analytical perspective allows us to explore the extent to which alternative combinations between politics, interests, and institutional layouts shape actor choices, thereby influencing the final characteristics of R&I instrument mixes. Here the conceptual pillars we hypothesize determining policy instrument choices are presented considering the national innovation systems features of the two countries. Different national R&I policy types are discussed together with alternative institutional organizational layouts and interest-intermediation strategies. Moreover, each section also includes a discussion on the working hypothesis driving the empirical analysis. Section 3.2 illustrates the data and methodology adopted for the analysis. Here data generation process is

C. Acciai, *Policy Design for Research and Innovation*, International Series on Public Policy,
https://doi.org/10.1007/978-3-031-36628-4_3

presented, together with the data sources collected during the analysis (more details on that can be found in the Appendix to the chapter). The final section of the chapter discusses the selection of the two cases and presents the comparative methodology applied in the analysis of their national innovation systems.

3.1 The Determinants of R&I Policy Design Choices

The following sections build on the theoretical framework illustrated in Chap. 2 and operationalize the hypothesized determinants of policy instrument choices in national R&I policy design. The first conceptual pillar focuses on the formulation of policy goals; it aims to understand how political priorities of the cabinet in office are translated into R&I policy strategies. Within this framework, the party politics orientation of the government is adopted as a proxy for a more general understanding of the political payoffs related to alternative R&I instrument choices.

The second conceptual pillar focuses on the institutional structures that allow for the design of policy strategies requiring coordination across sectors. Research and Innovation policies are characterized by competences often shared among different institutions; thus, policymakers have developed different strategies in the attempt to coordinate their functional responsibilities. Consequently, the ways policymakers foster synergies are influenced by how policymaking institutions are organized to coordinate their functional responsibilities. This affects how actors, in their interactions with institutional and organizational contexts, calibrate the characteristics of the inducement administered to policy recipients. Therefore, by illustrating the diachronic evolution of ministerial organizational layouts, it is possible to understand the role of the different ministries involved in national R&I policymaking, their organizational characteristics, their identity, and the role of their constituencies. This is closely related with the last component of our analysis: interest-intermediation strategies. Indeed, as ministerial organizational layouts investigate the internal coordination between policymaking institutions, the third pillar focuses on the external coordination between these actors and R&I stakeholders. Policy design literature has mostly approached the study of instrument selection from an "input perspective" (Howlett et al. 2020); hence, mostly focusing on the role of policymakers in choosing instruments. However, policy recipients have also increasingly acquired the power to nullify, or alter, the operation of policy instruments (Woodside 1986). Consequently, to overcome policy uncertainties and conflicts, policymakers have developed various intermediation strategies with the stakeholders, strategies that influence the way policy instruments are delivered to their recipients. The study describes alternative types of governing arrangements accompanying each instrument along its activation process and investigates how the different structures of external coordination influence the final instrument characteristics. In the following three sections of the chapter, these components are discussed in greater detail; furthermore, their capacity to determine instrument selection choices is investigated from a comparative perspective.

Nationally Consolidated R&I Policy Types

Innovation often represents a means to achieve broader political goals (e.g., grand societal challenges). Consequently, national R&I objectives become one of the instrumental paths through which governments attempt to reach these wider purposes, by means of meso-level innovation goals. How these objectives are to be translated into innovation-intensity targets, and the way these interventions should be balanced, represents a relevant political matter (Edquist and Borrás 2013).

As discussed in Chap. 1 in the historical overview of R&I policy rationalities, the logic behind the design of public actions in R&I has been driven by a multiple array of motivations. Starting from the highly top-down market-fixing logic of the World War II period, passing through the first-mover industrial advantage strategy of the post-war phase, to the systemic perspective of the National Innovation System, and finally to the instrumentality of R&I for tackling grand societal challenges.

Therefore, new policy initiatives have to deal with, and are bound to be affected by, long-term legacies emerging from earlier rounds of decision-making activities, and by the system of power distribution resulting from their consolidation over time (Edler et al. 2013; Flanagan et al. 2011; Hall 2010; Howlett 2009; Howlett and Mukherjee 2014; Linder and Guy Peters 1989). Policymakers attempt to make sense of the reality and to elaborate alternative policy strategies based on their interpretations of the problems at stake and on the room to maneuver provided by the legacy of previous choices. However, the extent to which they adopt one or the other rationality can only be assessed empirically (Béland 2009; Palier

2007). Therefore, it is useful to rely on the heuristic of the ideal types of technological governance theorized by Ergas (1987),[1] whose typology identifies two different R&I policy types: mission-oriented and diffusion-oriented.

Mission-Oriented Paradigm R&I Policy Types

Mission-oriented refers to strategies in which governments have a strong hand in the choosing of specific sectors for investments, by supporting large industrial efforts and organizing research through public institutions (Borrás and Seabrooke 2015). Here, policies are distinguished by a high degree of centralization in the determination of goals (Foray 2009; Foray et al. 2012). Public actors tend to play a pivotal role in the innovation process, leading the generation and exploitation of innovations (Chiang 1991). These approaches are also characterized by an emphasis on technology programs performed in-house (within public research organizations), mostly in the defense and aerospace sectors (Cantner and Pyka 2001; Chiang 1991; Ergas 1987). Public actors play a central role in the innovation process, in both leading the generation and exploitation of radical innovations, and in the creation of entirely new industries (Chiang 1991). Public investment is primarily concentrated toward large companies, which are perceived as the most appropriate actors, also infrastructurally, to develop innovation programs (Ergas 1987). Looking at this model of R&I policy with a policy instrument lens suggests how in this context it is more likely to prefer techniques of governance characterized by a direct control of government in both the choices and the enforcement of R&I policy instrument action. Indeed, this instrumentation relies on a more authoritative and less automatic mix to allow government a strong hand in the choice of the steering mechanisms characterizing national R&I strategies. The target of intervention is clearly defined and associated with technological objectives and future economic applications. Indeed France, with its national R&I system historically characterized by a tradition of state-financed industrial programs, consistently overlaps with the mission-oriented ideal type theorized by Ergas (1987).

[1] As pointed out by Cantner and Pyka (2001), the indirect classification method theorized by Ergas (1987) represents a crude insight into the technology policy of a specific country. Anyhow, given the comparative breadth of his classification and the richness of details it can embrace, it still represents a very useful heuristic tool to classify macro-level policy trends in R&I across different historical and political contexts.

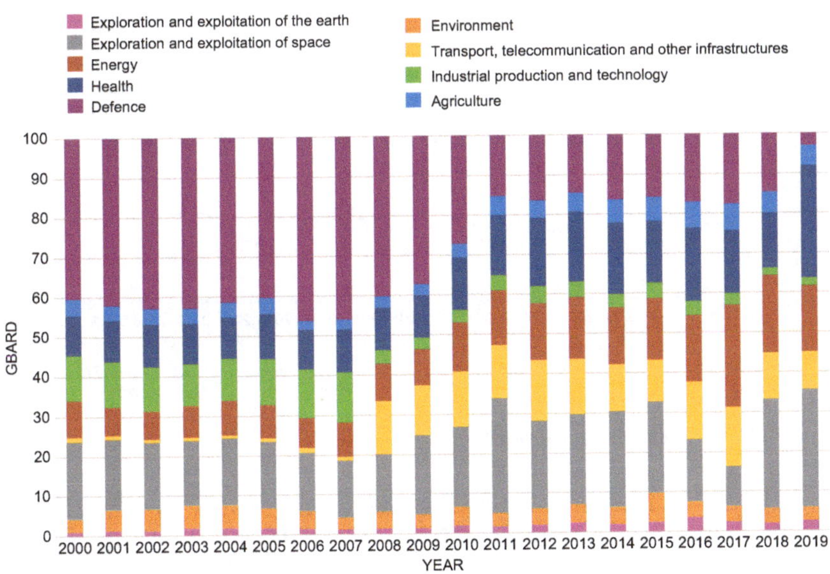

Fig. 3.1 GBARD statistics France

Government Budget Allocations on R&D—GBARD (Fig. 3.1)[2]—describe a national innovation investment portfolio where defense, agriculture, space exploration, and (to a smaller extent) energy represent a stable portion of public expenses.

These data reflect consolidated national preferences for specific R&D sectors of investments, which must be read in combination with the sector of performance mostly involved in the application of these budget allocations. French Gross Domestic Expenditures in R&D (GERD – Fig. 3.2) are dominated by private actors, with the government and the higher education sector following a parallel trend until 2005–2006, when the latter overtake in-house operators.

Therefore, in the French innovation system the State is closely involved in financing and executing R&D in the civilian sector. This characterizes a fairly centralized model able to ensure stability with regard to technological choices (OECD 2014). These features mirror some of the constitutive components of the mission-oriented type (Berger 2016), distinguished by

[2] Source for GBARD and GERD data: OECD. Data elaborated by the author.

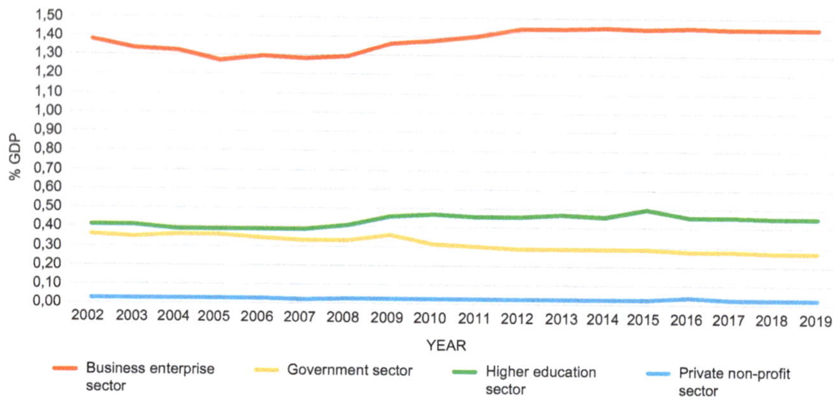

Fig. 3.2 GERD statistics France

the choice to invest in specific technological sectors associated with national objectives and strategies for economic development.

Diffusion-Oriented R&I Policy Types

This policy type describes strategies aimed at upgrading knowledge competences across different sectors of R&I governance systems. The main purpose of public intervention is widely defined as increasing overall national economic capacity, by concentrating public resources on different processes like investments in economic infrastructures, technology transfer, cooperative R&D, creation of university, and industry links (Cantner and Pyka 2001; Ergas 1987). Public interventions are primarily focused on the acquisition, diffusion, and assimilation of technologies for national industries (Chiang 1991), and governments endow heterogeneous sectors to approximately the same extent. Research targets are broadly defined, as well as the spectrum of industrial and technological recipients. Their instrumentation relies on less authoritative and more automatic instrument mixes. Policy decisions tend to be taken in a decentralized manner, governmental agencies play a limited role in the implementation, and these tasks are delegated to either industrial associations or cooperative research organizations (Ergas 1987). Indeed, national R&D budget allocations are consistently more scattered around different sectors, especially in comparison with France (Fig. 3.3).

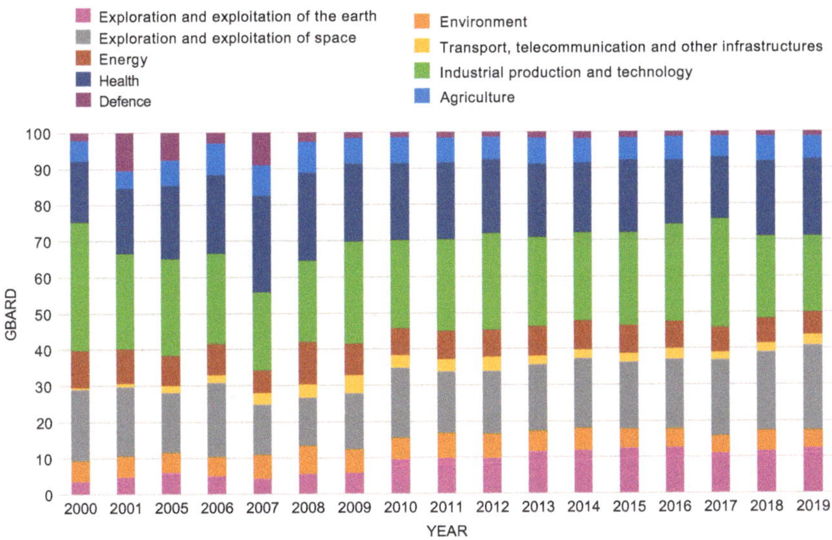

Fig. 3.3 GBARD statistics Italy

There is some degree of stability in sectors like health and industrial production; however, it is necessary to bear in mind these tables show relative values (the shares of investments over the total GBARD expenses). Consequently, Italy scores a generalized, and substantial, lower level of R&D expenditures compared to France. This becomes evident when moving to Fig. 3.4 where, despite the substantial increase in the business and enterprise sector of performance, government R&D expenditures are comparatively lower. Consequently, the combination between the increasing relevance of business actors and the low level of in-house public investment is consistent with the R&I diffusion-oriented policy type, which tends to rely on innovative activities directly undertaken by R&I performers. The national industrial system is mostly oriented toward incremental innovations and the dominant instrument mix is particularly centered on providing SMEs with the necessary instruments to design their innovation strategies autonomously (Evangelista 2007; Gallo and Silva 2006). Consequently, the country suffers from an inherent fragmentation and duplication of instruments, coupled with scarcely effective monitoring, and uncertainties related to the implementation process and the availability of resources (Potì and Reale 2011).

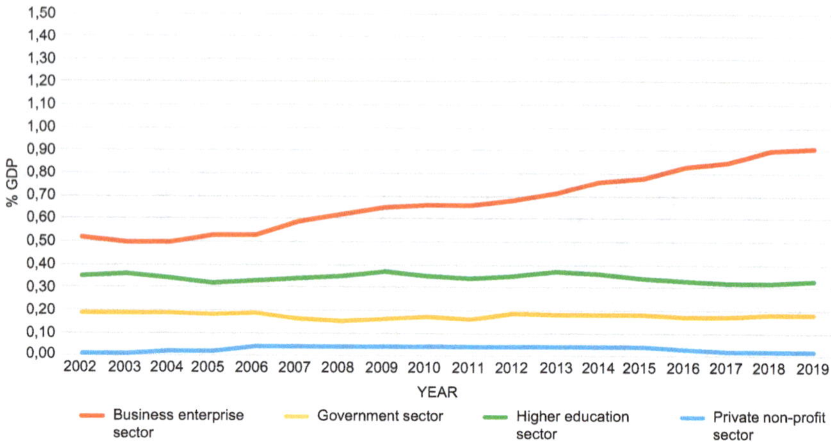

Fig. 3.4 GERD statistics Italy

Mission- and diffusion-oriented strategies represent different paradigmatic models of R&I policy according to which policies are designed and organized. This typology shows how, despite decision-makers being free to choose the instruments they prefer or those they find more appropriate according to their evaluations, these choices can be influenced by the legacy of dominant instrument mixes and related institutionalized practices of action (Bouwma et al. 2017; Cashore and Howlett 2007; Flanagan et al. 2011; Gassler et al. 2008; Hall 2010; Howlett 2009; Howlett and Mukherjee 2014; Ringeling 2005). Therefore, long-term state and societal preferences can constrain politically available design choices (Jungblut 2015). Indeed, path dependency forces have been shown to have a different influence according to the partisan acceptance of dominant R&I policy types (Acciai 2021). However, policymakers tend to avoid being politically related with instrument mixes that are not associated with their political framings (Voß and Simons 2014). Consequently, we would expect that *different policy instruments would be selected in relation to different cabinets' party politics orientations.* Families of instruments embed different typologies of authoritative relationships between decision-makers and policy recipients. These represent different means-ends rationalities of the inducement implemented by public action. Similarly, the different instrumentation of diffusion and mission-oriented R&I policy types reflect well-established preferences on what governments should do, and how they should make it happen. Consequently, since political competition is about the control over resources and their management, *we expect to identify different trends in the distribution of government choices for R&I policy instrument features across countries characterized by different traditions in R&I policy types.*

Ministerial Organizational Layouts

Policy design takes place within specific institutional structures. This system of norms and rules influences the way actors build their expectations regarding others' actions (John 2012), thereby producing regularities on aggregate behaviors. Institutions influence the functioning of coordination, affecting information exchange and agents' strategies (Peters 2000) according to how powerful actors are distributed (Borrás and Edler 2015; Lepori 2011). The relevance of this for R&I policies is striking. By their nature, R&I policies are characterized by competences often shared among different institutions (at national or sub-national levels), defining a multi-level and multi-actor governance structure (Chou et al. 2017; Edler et al. 2016; Edler and Fagerberg 2017). Consequently, policymakers have developed different strategies in the attempt to coordinate their functional responsibilities. This has often produced a blurring of competences between executive departments, affecting the way institutions with complementary competences in the sector are organized (Edler et al. 2016; Edler and Fagerberg 2017; Perry and May 2007). Hence, it is necessary to differentiate between formal and informal types of institutions. In the former, rules and procedures are created, and enforced, through channels widely accepted as official (Helmke and Levitsky 2004:727); while in the latter, rules are sustained and enforced by the actors populating those formal institutions. Institutions add stability to the political system, but due to the fact that they interact with actors, interests, and ideas (Blyth 2003), they can also generate unintended effects. Therefore, formal and informal institutions assist one another and together represent the backbone of governance arrangements (Borrás 2009; Peters 2012). The scholarship provides different conceptualizations of these practices. Braun (2008b) calls them external and internal coordination practices, focusing on the type of ministerial coordination necessary to guarantee coherence in public action, whereas Verhoest and Bouckaert (2005) speak about horizontal, or vertical, specialization strategies to align departmental structures. Both typologies aim at describing how national governments frame the

Internal Specialization Patterns	*External Specialization Patterns*
Overall competences related to R&I have been centered around one *superministry* (Braun 2008a), who has acquired all the capabilities and decisional leadership over the entire policy sector. The differentiation of responsibilities take place within a unique organizational structure in a vertical manner among departments or directorates. Attempts at coordination hinge upon intra-organizational bargaining, hence transaction costs between different areas are internalized. Different tasks across the policy cycle are split-up among departments within the same organization, and issues of administrative coordination can be tackled in the shadow of ministerial hierarchy.	Policy responsibilities are decoupled among functionally homogeneous organizations, and specialization is organized horizontally among peers. Ministers attempt to coordinate externally, by means of different types of interdepartmental bargaining (Braun 2008b). In many national examples these have been supplemented by the creation of platforms for interministerial dialogue or common budgeting procedures. However, decision-making venues that do not have their own organizational identity tend to be an arena for exchanging information, dependent upon the political will of singular ministers (Braun 2008a) rather than an effective policymaking laboratory.

Fig. 3.5 Typology of institutional layouts

political priorities to be tackled, in terms of formal institutional structures and the resulting coordination practices necessary to fuel policy design. This research project adopts the classification of formal institutions proposed by Braun (2008a), while also integrating some insights emerging from the analysis of Verhoest and Bouckaert (2005), and identifies two types of ministerial organization patterns (Fig. 3.5).

Both types of institutional layouts experience transaction costs related to the need for coordination between actors with shared responsibilities. Indeed, policymaking institutions designing boundary-spanning policies (R&I) experience issues with both different systems of competence division (Pelkonen et al. 2008) and with the mechanisms of habituation and socialization characterizing organizational structures (March and Olsen 1997; Merton 1938). These frictions influence how authority is exercised, and they both require striking a balance between the need to design an all-encompassing policy strategy (coordination) and the calibration of instrument mixes to the conditions of specific contexts (specialization). For these reasons, we expect that (formal) institutional structures will influence the ability of different ministries (or departments) to collaborate. Hence, *countries undertaking similar ministerial specialization strategies (formal institutions) will likely adopt similar coordination practices (informal institutions).*

The Configurations of Interest Intermediation Strategies and their Actors

In policymaking activities requiring technical competencies and skills decision-makers cannot provide by themselves (like R&I), public actors are likely to delegate certain tasks to external agents. Consequently, policy recipients have increasingly acquired the power to nullify, or alter, the operation of policy instruments (Woodside 1986). This two-way bond reflects the inability of contemporary governments to move unilaterally, neglecting preferences and resources of other social actors involved in policymaking (Bressers and O'Toole 2005; Fraussen 2014; Smits et al. 2010). Policy tools embed the aspects of a policy intended to motivate a target population to comply with a policy or utilize policy opportunities (Schneider and Ingram 1993:338). Therefore, their features are molded to account for the characteristics of recipients.

In order to institutionalize conflict and to overcome uncertainties in their dialogue with target populations, policymakers have developed various intermediation strategies (Braun 1993; Van der Meulen 1998). Across national contexts, specialized R&I agencies are appointed to the management and development of policy strategies, with the task of translating political guidelines, or thematic priorities, into actions (Nauwelaers and Wintjes 2008). Public agencies are intermediary actors located in-between societal spheres, with the task of institutionalizing social interfaces and creating common contexts for action (Hartmann and Kjaer 2015). Their development and diffusion reflect New Public Management (NPM) principles, according to which national bureaucracies are increasingly relying on external bodies and committees to formulate and implement their policies (OECD 2005). In the specific case of R&I policy, NPM has indeed introduced greater decentralization, contract management, privatizations, PPPs, and, more generally, the dispersal of power to quasi-state and non-state actors (Borrás 2009; Flanagan et al. 2011). However, public agencies in R&I sectors acquire various organizational characteristics, according to existing structures of interest intermediation between policymakers and target groups.

In order to operationalize such a variety, this research builds upon the typology of Arnold and Boekholt (2003), as illustrated in Fig. 3.6.

The first type of policy agency is the mono-principal agency, an intermediary actor politically bound to only one minister, that can simultaneously interact with one or more recipients. The second type is multi-principal

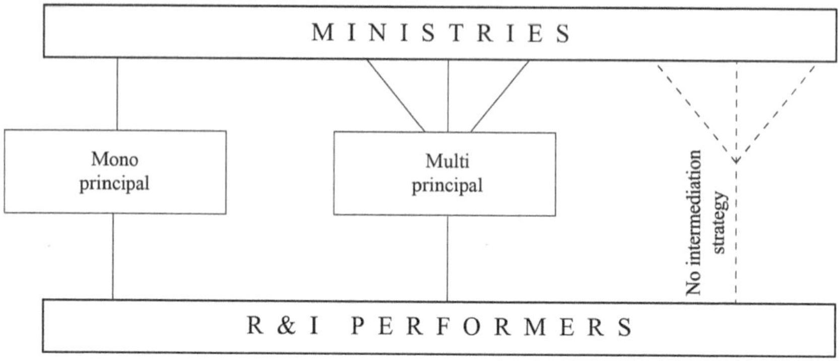

Fig. 3.6 The variety of intermediation strategies

agency. This represents a broader management platform that can be employed by different ministers for different target populations and can also be used as a platform to define coordination mechanisms among ministers (multi-principal). Finally, there are instances where governing actors do not make use of any intermediary actor to stipulate delegation contracts with different target groups. In these cases, policymakers engage in dialogue directly with the world of performing agents based on their responsibilities. Consequently, R&I policies are steered directly by the minister (or ministries), under the supervision of the Parliament (OECD 2014:125). As discussed in Chap. 1, establishing intermediary actors is a strategy aimed at externalizing and de-politicizing conflict between policymakers and R&I performers by making use of technical and managerial motivations. Consequently, the variety of policy competences delegated to these actors[3] will likely influence the relationship between policymakers and with R&I performers.

France and Italy adopted different approaches in their interest-intermediation strategies. The first case demonstrates a consolidated practice of relying on different types of public agencies like, for example, the *Agence nationale de la recherche* (National Research Agency, ANR) for financing public research and the OSEO agency for the support of industrial innovation projects. Interestingly, an increasing reliance on public intermediary agencies started exactly during the period under

[3] These can be research councils, ministerial agencies, or technological agencies, depending on the national specificities. Despite their different areas of interests, all these actors share the role of intermediary between the principal and the agent.

investigation. Italy is characterized by direct intermediation practices, usually between each minister and their constituencies. The country had some short-term experiences with public agencies; however, these actors had a short life, and a minor role, in interest-intermediation practices.[4]

All the typologies presented in Fig. 3.6 are not immune to the fragmentation of ministerial powers, to the extent that internal policymaking fragmentation can be directly transposed into the relationship with R&I performers. Since ministers play the role of coordinating objectives and instruments (Chung 2013), when they experience organizational fragmentation it is more challenging to design a holistic strategy for publication. Indeed, each policymaking unit will likely pursue their individual strategies, increasing the risk of redundancies and internal decisional deadlocks. Therefore, the higher the fragmentation of policymaking duties, the more complicated it will be to design a clear inducement that can shape target population behaviors. On the other hand, skillful constituencies of R&I performers can exploit institutional stalemates to their advantage, to see their interests represented. Therefore, it is likely to expect that *the higher the level of institutional fragmentation among principals, and the higher the capacity of R&I performers to behave as a political constituency, the more likely it will be to select a more automatic instrument mix.* This is because instruments with automatic delivery components require less public engagement in their activation, while relying more heavily on the autonomous uptake of recipients and stakeholders involved in their implementation.

3.2 Research Design and Methods

The methodology adopted for this research relies on a combination of different sources (budgetary documents, experts, and stakeholders' interviews) and analytical techniques. Similar to the exploratory framework of analysis, the research design is also distributed across two levels. A macro level, characterized by the investigation of national R&I partisan politics (political orientations of the cabinet in office) and a micro level, focused on the interactions between ministerial organizational layouts (institutions) and interest-intermediation strategies (interests) as described in Fig. 3.7.

[4] Since both France and Italy have the same ministerial organization strategy in R&I (external specialization pattern), the influential role that institutional organizational layouts play can be "held constant," hence allowing for a closer examination of policy recipients.

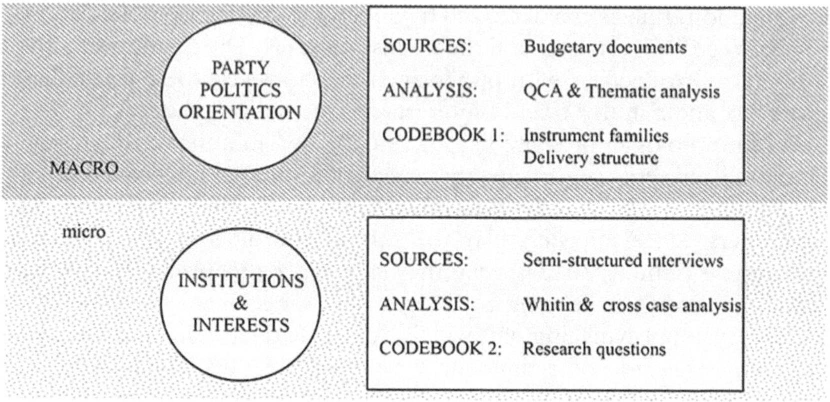

Fig. 3.7 The methodological strategy

The research design combines document analysis (macro) with semi-structured and expert interviews (micro). Each technique represents a piece of a broader picture intended to capture different perspectives of R&I policy design while validating the findings by complementing and providing additional coverage to each other (Dür 2008). The macro analytical level makes use of policy documents,[5] providing information on national economic R&I policy planning, together with cabinet commitments on the allocation of resources. These sources provide a perspective on the R&I means-end rationalities of different governments, allowing us to reconstruct the architecture of choices from the perspective of decision-makers (Giest 2016). Data are analyzed and interpreted through a unidimensional codebook, relying on the instrument family and delivery components dimensions of the proposed treatment of policy instruments (Sect. 3.2). Specifically, each dimension represents mutually exclusive categories—of authority and automaticity, respectively—which are associated with different data points (portions of text). Then, combining Qualitative Content Analysis (QCA) (Schreier 2012) and Thematic Analysis

[5] The total number of budgetary policy documents analyzed is equal to 32, evenly distributed across the two cases. The document analyzed for the French case was the yellow appendix ("*Annexes Jaunes*"), while for Italy the Economic and Financial Planning Document ("*Documento Programmatico di Economia e Finanza*"). So, due to similar institutional characteristics of these documents, we have been able to ensure a high degree of comparability and scope-similarity of the information analyzed.

methodologies (Boyatzis 1998), the relationships between latent themes are identified. This exercise consists of matching portions of documents containing statements about the selection of an instrument (or on the conservation of existing ones) with the categories of our codebook (for more details on this procedure and the codebook please refer to the Appendix). Then, each document is assigned to a cabinet period in order to investigate the diachronic evolution of R&I instrument choices across different government periods.[6]

Differently, institutions and interests stand at a lower level of abstraction, and they are analyzed jointly since they illustrate the way policymakers and policy takers interact, ultimately being "two sides of the same process." The data generation process relies on a set of semi-structured interviews with experts and stakeholders.[7] Interviewees were identified according to criterion-based (or purposive sampling) techniques (Ritchie and Lewis 2003). Experts were identified as those individuals with the necessary knowledge to structure collective problems (Dente 2014) but who were not directly involved in the policymaking process as representatives of political interests.[8] Interviewing actors with no direct stakes in the process, but with the necessary knowledge to understand its dynamics, illustrates how R&I issues are structured within different national contexts. Differently, stakeholder interviews involved those actors directly engaged in R&I policymaking and investigated their cognitive perceptions and values. This second pool of interviewees was created according to a theoretical sampling technique: by choosing actors according to their roles in national R&I policymaking.

The total sample consisted of one actor per each national organization involved in the R&I governance, as a proxy for the perspective of their organization of affiliation.[9] This resulted in a total of 22 interviews, evenly

[6] The analysis employed the computer-aided software Nvivo12.

[7] Some of the questions included in the questionnaires were kept constant, while others have been differentiated according to the role played by the interviewees in their respective national R&I governance.

[8] Therefore, experts represent researchers, or scientists, who worked in academia or public research organizations and possess an education title or consolidated experience in a specific field (Rimkutė and Haverland 2015).

[9] The sampling techniques maintained a high degree of comparability both within cases (in terms of coverages of the organizations involved in the national R&I sector) as well as across cases, since many interviewees played the same roles within their own national contexts. Allowing also for (when needed) an iterative process between sample selection, fieldwork, and analysis (Ritchie and Lewis 2003).

distributed across the two cases,[10] which have been transcribed and analyzed through a methodological approach to qualitative data analysis inspired by within and cross-case analysis techniques (Miles and Huberman 1994). Each portion of text was assigned to a set of unidimensional concepts, in accordance with the four lines of investigation of this study (see Fig. 3.2 in the Appendix).[11] This allowed us to systematically interpret, and describe, the meanings of different sources (Schreier 2012); and to identify similarities and differences between actors and their organizations, both within and across countries (Burns 2010; Chmiliar 2010; Mathison 2005; Miles and Huberman 1994). Finally, to reconstruct national policy-making processes, these data were supplemented by secondary literature.

Case Selection

Knowledge infrastructures evolve in combination with national political and institutional systems (Edler and Fagerberg 2017). Similarly, countries differ in their demand of skills, finance, and political strategies, as demonstrated by the comparative study of Nelson (1993). In line with the aforementioned evidences, the current study adopts a country-based level of analysis since most public policies directly affecting the characteristics of R&I strategies are designed at the national level (Edquist 2001). France and Italy have been selected as exploratory cases for analysis because of the high degree of variance they display in the theoretical factors of interests we hypothesize are influencing R&I instrument choices (Seawright and Gerring 2008). According to a typological logic (Alexander and Bennett 2005), the research investigates different combinations of analytically relevant factors to shed light on the causal pathways explaining policy design and instrument choices. The case-study investigation diachronically reconstructs different national historical sequences (Rueschemeyer and Stephens 1997) by tracing national R&I policy design processes in parallel with the evolution of national instrumentation.

France and Italy show opposite consolidated R&I policy types of practices (Acciai 2021) with respect to the paradigmatic models of R&I policies. In the case of France, these practices consistently overlap with the

[10] For further details on the type of interviewees, please refer to Table 3.1 in the Appendix of this chapter.

[11] These are introduced in Chap. 1, while a copy of the codebook is presented in the appendix of this chapter (Table 3.2).

mission-oriented R&I policy type, while Italy displays many similarities with the diffusion-oriented model. These differences are pronounced also in their interest intermediation practices. In the French case, these strategies rely on a differentiated set of public agencies, like the National Research Agency (ANR) for funding public research, or the OSEO agency which supports industrial innovation projects. Differently, Italy is characterized by direct interest-intermediation practices, usually established between a functional minister and its related constituencies. The country had some short-term experiences with R&I public agencies, but these actors had a short life and especially played a minor role in the national interest-intermediation strategy. Despite these fundamental differences, France and Italy do share some (formal) similarities in their institutional specialization patterns. In both cases, R&I policies' competences lie under the remit of the Ministry for Higher Education and Research, and the Ministry with economic competences (called Minister of Economic Development, in the Italian case, and minister of economy and finance in the French case). Since ministerial organizational choices are not strictly determined by law, institutional organizational strategies tell us how national governments think about policy problems and alternative strategies to tackle them (Peters 1998a). Therefore, this evidence suggests some similarities in the national organization of R&I policy tasks in France and Italy, hence supporting a baseline for their comparison.

Given the exploratory nature of the study, and due to the ongoing effort in classifying and describing an R&I instrument mix, the case-selection strategy accounted also for possible multiple sources of extraneous variance: for example, their history, culture, and societal developments. Indeed, France and Italy share some similarities in the fundamental characteristics of their national R&I governance. These countries have similar national economic size,[12] they display similar strategies of competence distribution among actors—namely a context where specialized research and funding are (mainly) kept as central competences—while competences in lower levels of education, cluster policies, and technology transfer (with their related incubation activities) tend to be devolved at the regional level. They also share very similar administrative traditions, to the extent that the Italian administrative system is, in many respects, one of the most

[12] They are respectively the second and third largest European economies in terms of contribution to the European Gross Domestic Product, right after Germany https://ec.europa.eu/eurostat/web/products-eurostat-news/-/DDN-20200508-1 (last access, April 2021).

Napoleonic outside of France (Spence 2014:104). Therefore, from the perspective of R&I policies, these countries are two apparently similar and binary systems; however, within this similarity, there are also very different approaches characterizing national policymaking dynamics. The time frame of the analysis covers approximately 12 years of R&I policy design: an appropriate time span to examine policy change (Sabatier 1986). To conclude, the characteristics of our outcome of interest—R&I instrument mixes have been investigated in parallel with the case-study analysis. Official databases and documents typically do not capture policy mix characteristics, or their operationalization, thus requiring first-hand data collection and interpretation (Rogge and Reichardt 2016). This analysis proved also to be a necessary exercise to test the validity of the instrument classification typology proposed by our research.

Appendix

Multi-Tiered R&I Governance Structure: A Piloting Exercise on Interviewees

Research and Innovation policies have been included within the European Open Method of Coordination (OMC) strategy since the earliest stages of its development. By its very nature, the OMC represents a set of recommendations and guidelines aiming at balancing member states' decision-making autonomy with a centrally imposed "one size fits all" European strategy (Buchs 2007; Pochet and de la Porte 2002). It works through coordinated agreements between member states on benchmarks and indicators to measure best practices, which should then be transposed into national and regional policies. Research and Innovation policies were included in European policy documents from the early 1990s, starting with the Green Paper on Innovation and the First Action Plan for Innovation in Europe (Seravalli 2009). After the adoption of the Lisbon strategy (2000), there has been a clear focus on developing a knowledge-based economy in Europe through investments, especially in R&I. This led to a stronger focus from European Structural Funds for supporting regionally based R&I policies, which became a central element of supranational strategies for national competitiveness (Reillon 2015). To support the diffusion of innovation-oriented policy actions among member states, European regulations have changed accordingly. One example is the reform of the clauses granting exemptions on competition policy regulation concerning R&D agreements. Indeed, the EU is said to have moved away from a legalizing approach to competition law, toward an economic approach based on the analysis of market impact and the potential market dominance of large R&D alliances (Borrás 2009:6).

With the modification of the Treaty on the Functioning of the European Union by the Treaty of Lisbon in 2007, research policy was moved from a supportive to a shared competence, hence enabling the EU to adopt binding acts (Reillon 2015:2). Furthermore, the European strategy for growth presented in March 2010 (EU 2020 Strategy for smart, sustainable, and inclusive growth) further increased the relevance of R&I for supranational policy strategies. In parallel with the growing awareness of the relevance of localized opportunities for innovation development, European programs started to support regional development through the promotion of innovation-related interventions (Prange 2008). This is demonstrated by the increase in the share of funds allocated to innovation-related activities in cohesion policies (Seravalli 2009); as well as in the framework of the new regulation on European Regional Development Funds (ERDF).[13] Indeed 50–80% of ERDF funds were supposed to be dedicated to, at least, two of the following objectives: strengthening research, technological development, and innovation; enhancing access to ICT; competitiveness of SMEs, of the agricultural sector, and of the fishery/aquaculture sector; and supporting the shift toward a low-carbon economy in all sectors (Reillon 2015). Within this framework, national governments can leverage European instruments to design R&I policy (Prange 2008) and the shape of (ERDF-driven) regional innovation policies represent a partial revision of the classical European cohesion policies toward a more horizontal-oriented approach (Seravalli 2009). Nonetheless, despite the role the European Union could play in pushing toward regionalization in some policy areas, regional initiatives remain to a large extent related to intra-national framing of problems (Olsson and Åström 2003). Consequently, European institutions play a role in the framing of European R&I strategies, as well as in identifying the objectives related to specific structural funds. However, also national governments take part in European decision-making processes, thereby ultimately influencing the framing of European objectives.

[13] Regulation EU No 1301/2013 of the European Parliament and the European Council 17/12/2013.

Given the interplay between national and supranational policymaking capacity in shaping national R&I policy objectives, we decided to run a pilot exercise during our interviews to understand the perceptions of the interviewees across our two cases. Consequently, each interview questionnaire included a set of standardized questions (the same for different "types of actors" across the two cases) on the role of the European Union and regional actors for national R&I policymaking. According to the interviewees, the major European-driven constraints national policymakers are facing in designing R&I strategies are state aid regulations. These rules are perceived as reducing their freedom for direct intervention in different industrial sectors (INTERVIEWEE 8, 5, 12, 21). Moreover, there clearly emerges a national attitude toward the emulation of some successful European experiences at the national level, like the European Research Council (INTERVIEWEE 2, 13, 15, 14, 3, 12). There is also a widespread agreement concerning the active role that national policymakers play in the definition of the European R&I agenda (INTERVIEWEE 7, 10, 2). Finally, European objectives are perceived to be broadly framed, hence leaving enough room for maneuver in the identification of national goals (INTERVIEWEE 22, 11, 2, 18).

In addition to the European perspective, a regional one has also been considered due to increasing evidence that the governance of innovation policies has been more and more oriented toward a multi-level governance dimension (Flanagan et al. 2011; Perry and May 2007; Tödtling and Trippl 2005, 2012). In both France and Italy, regions have frequently complemented R&I national measures, particularly in the framework of competitive clusters. The involvement of regional and local authorities has sometimes increased the complexity of national policy strategies, triggering administrative confusion, and the layering of similar instruments across different territorial levels (INTERVIEWEE 14, 10). Regional governments are perceived as complementary funders of national R&I interventions (INTERVIEWEE 16), as well as the best administrative level to implement specific national policies (INTERVIEWEE 17, 5, 2).

In both cases of regional and European roles for national policy design, interviewees suggest how, despite the importance of these dimensions, these actors do not seem to interfere in national R&I policy choices. This is because R&I tends to be a sector which could heavily contribute to a country's economic growth; therefore, the scale of intervention is often kept at the national level (Tamtik 2016). The interaction and networks between actors are shaped by national institutions; and national governments play a central role in their establishment (Chung 2013:1053). That is the reason why, ultimately, a country-based level of analysis is the most suitable to understand the underlying dynamics of R&I policy design.

Therefore, the political battle over the allocations of R&I resources takes place inside the political space defined by internationally accepted models of R&I policies while delineating specific country policy profiles (Thèves et al. 2007). National policymakers maintain the lead in national R&I policy choices, since the adoption of certain European procedures and policy instruments is considerably influenced by national policy legacy, context, and interest structures (Capano et al. 2012; Héritier and Lehmkuhl 2011).

Table 3.1 List of interviewees

INTERVIEWEE	*Type*	*Role in the national context*	*Country*
INTERVIEWEE 1	Stakeholder	Representative of universities	Italy
INTERVIEWEE 2	Expert	Advice to government in R&I issues	Italy
INTERVIEWEE 3	Expert	Advice to government in R&I issues	Italy
INTERVIEWEE 4	Expert	Academic expert on national R&I	Italy
INTERVIEWEE 5	Stakeholder	Ministry of Economic Development	Italy
INTERVIEWEE 6	Expert	Representative of National Public Agency in R&I	Italy
INTERVIEWEE 7	Stakeholder	Representative of national PROs	Italy
INTERVIEWEE 8	Experts	Advisor on R&I issues to stakeholders	Italy
INTERVIEWEE 9	Stakeholder	Ministry of Higher Education and Research	Italy
INTERVIEWEE 10	Stakeholder	Confindustria (confederation of industrials)	Italy
INTERVIEWEE 11	Stakeholder	Conference of Italian University Rectors	Italy
INTERVIEWEE 12	Expert	Academic expert on national R&I	France
INTERVIEWEE 13	Expert	Advice to government in R&I issues	France
INTERVIEWEE 14	Expert	Advice to government in R&I issues	France
INTERVIEWEE 15	Expert	Former director of departments in different R&I-related ministries ("insider view")	France
INTERVIEWEE 16	Expert	Advice in R&I issues to different (mainly private) stakeholders	France
INTERVIEWEE 17	Stakeholder	Ministry of Higher Education and Research	France
INTERVIEWEE 18	Stakeholder	Conference of University Presidents	France
INTERVIEWEE 19	Stakeholder	Representative of CNRS	France
INTERVIEWEE 20	Stakeholder	MEDEF (confederation of industrials)	France
INTERVIEWEE 21	Stakeholder	Ministry of Economy, Finance and Industry	France
INTERVIEWEE 22	Stakeholder	National Research Agency	France

Table 3.2 Codebook and description of the contents

Nodes	*Exemplifying Content Included*
Cabinet politics and R&I policymaking	(Interviewees' perception over) the influence of cabinet party politics on R&I policy design choices (e.g., different policy choices across different cabinet periods, some governments were promoting specific combinations of R&I instruments)
The influence of history and path dependency forces on R&I policymaking	(Interviewees' perception over) the extent to which previous policy design choices/practices limit or enhance current R&I policy design choices
Ministerial structures (formal institutional specialization patterns)	(Interviewees' perception over) how and why the division of R&I competences takes place across ministers. And the way different cabinets have dealt with it
Ministerial interaction patterns (informal institutional specialization patterns)	(Interviewees' perception over) which types of (in)formal coordination and collaboration practices take place between ministers and the extent to which these affect the functioning on policy design activities
Target group relationship	(Interviewees' perception over) the ways different private and public stakeholders interact within the national R&I governance space
Target group lobbying	(Interviewees' perception over) the ways R&I stakeholders interact with decision-makers and some successful examples of dialogue
European role	(Interviewees' perception over) how supranational policy choices influence national R&I policy design choices
Region role	(Interviewees' perception over) how regional policymaking influences national R&I policy design choices

Bibliography

Acciai, Claudia. 2021. The Politics of Research and Innovation: Understanding Instrument Choices in Complex Governance Environments – the Case of France and Italy. *Research Policy* 50 (9): 104254.

Acciai, Claudia, and Giliberto Capano. 2021. Policy Instruments at Work: A Meta-Analysis of Their Applications. *Public Administration* 99: 118–136.

Alexander, George L., Andrew, Bennett. 2005. *Case Studies and Theory Development*. Cambridge: MIT Press.

Arnold, Erik. 2007. CREST 3 % OMC Third Cycle Policy Mix Peer Review Country Report France.

Arnold, Erik, and Patries Boekholt. 2003. Research and Innovation Governance in Eight Countries: A Meta-Analysis of Work Funded by EZ (Netherlands) and RCN (Norway).Technolopolis.

Banfi, A., and G. Viesti. 2016. Il Finanziamento Delle Università. In *Università in declino. Un'indagine sugli atenei da Nord a Sud*, ed. G. Viesti, 277–304. Donzelli. Roma.

Baumgartner, Frank R., and Bryan D. Jones. 1993. *Agendas and Instability in American Politics*. Chicago: University of Chicago Press.

Bayerlein, Louisa, Christoph Knill, and Dionys Zink. 2021. The Concept of Administrative Styles. In *The Routledge Handbook of Policy Styles*, ed. M. Howlett and J. Tosun, 153–164. London: Routledge.

BCG & CM INTERNATIONAL. 2008. *Evaluation Des Pôles de Compétitivité, Synthèse Du Rapport d'évaluation.*

Béland, Daniel. 2009. Ideas, Institutions, and Policy Change. *Journal of European Public Policy* 16 (5): 701–718.

Béland, Daniel, and Michael Howlett. 2016. The Role and Impact of the Multiple-Streams Approach in Comparative Policy Analysis. *Journal of Comparative Policy Analysis: Research and Practice* 18 (3): 221–227.

Bergek, Anna, Marko Hekkert, Staffan Jacobsson, Jochen Markard, Björn Sandén, and Bernhard Truffer. 2015. Technological Innovation Systems in Contexts: Conceptualizing Contextual Structures and Interaction Dynamics. *Environmental Innovation and Societal Transitions* 16: 51–64.

Berger, Suzanne. 2016. Reforms in the French Industrial Ecosystem.

Bersani, Pier Luigi, Alfonso Lelio, and Letta Enrico. 2004. *Viaggio Nell'economia Italiana*. Donzelli.

Blyth, Mark. 2003. Structures Do Not Come with an Instruction Sheet: Interests, Ideas, and Progress in Political Science. *Perspectives on Politics* 1 (4): 695–706.

Bologna, Federica, Angelo Di Iorio, Silvio Peroni, and Francesco Poggi. 2021. *Do Open Citations Inform the Qualitative Peer-Review Evaluation in Research Assessments? An Analysis of the Italian National Scientific Qualification.*

Borrás, Susana. 2009. The Widening and Deepening of Innovation Policy: What Conditions Provide for Effective Governance? In *Papers in Innovation Studies 2009/2, Lund University, CIRCLE - Center for Innovation, Research and Competences in the Learning Economy.*

Borrás, Susana, and Jakob Edler. 2015. *The Governance of Socio-Technical Systems: Explaining Change*. Cheltenham: Edward Elgar Publishing.

Borrás, Susana, and Charles Edquist. 2019. *Holistic Innovation Policy*. Oxford University Press.

Borrás, Susana, and Leonard Seabrooke. 2015. Sources of National Institutional Competitiveness: Sensemaking in Institutional Change.

Bouwma, Irene, Bas Arts, and Duncan Liefferink. 2017. Cause, Catalyst or Conjunction? The Influence of the Habitats Directive on Policy Instrument Choice in Member States. *Journal of Environmental Planning and Management* 60 (6): 977–996.

Boyatzis, Richard E. 1998. *Transforming Qualitative Information: Thematic Analysis and Code Development*. Thousand Oaks: Sage Publications.

Braun, Dietmar. 1993. Who Governs Intermediary Agencies? Principal-Agent Relations in Research Policy-Making. *Journal of Public Policy* 13 (02): 135.

———. 2008a. Lessons on the Political Coordination of Knowledge and Innovation Policies. *Science and Public Policy* 35 (4): 289–298.

———. 2008b. Organising the Political Coordination of Knowledge and Innovation Policies. *Science and Public Policy* 35 (4): 227–239.

Bressers, H.A., and L. O'Toole. 2005. Instrument Selection and Implementation in a Networked Context. In *Designing Government: From Instruments to Governance*, ed. P. Eliadis, M. Hill, and M. Howlett, 132–153. Montreal: M.-Q. U. Press.

Bressers, Hans, and Laurence O'Toole. 1998. The Selection of Policy Instruments: A Network-Based Perspective. *Journal of Public Policy* 18: 213–239.

Buchs, Milena. 2007. *New Governance in European Social Policy: The Open Method of Coordination*. Palgrave Macmillan.

Burns, J.M.C. 2010. Cross-Case Synthesis and Analysis. In *Encyclopedia of Case Study Research*, ed. A.J. Mills, G. Durepos, and E. Wiebe, vol. 1. Thousand Oaks: SAGE Publications.

Cantner, Uwe, and Andreas Pyka. 2001. Classifying Technology Policy from an Evolutionary Perspective. *Research Policy* 30 (5): 759–775.

Capano, G., J.E. Rayner, and A.R. Zito. 2012. Governance from the Bottom up: Complexity and Divergence in Comparative Perspective. *Public Administration* 90 (1): 56–73.

Capano, Giliberto, and Elisabetta, Gualmini. 2011. Le Pubbliche Amministrazioni in Italia.

Capano, Giliberto. 2008. Looking for Serendipity: The Problematical Reform of Government within Italy's Universities. *Higher Education* 55 (4): 481–504.

Capano, Giliberto, and Andrea Lippi. 2017. How Policy Instruments Are Chosen: Patterns of Decision Makers' Choices. *Policy Sciences* 50 (2): 269–293.

Capano, Giliberto, Andrea Pritoni, and Giulia Vicentini. 2019. Do Policy Instruments Matter? Governments' Choice of Policy Mix and Higher Education Performance in Western Europe. *Journal of Public Policy* 40(3): 375–401.

Capano, Giliberto, Marino Regini, and Matteo Turri. 2016. *Changing Governance in Universities: Italian Higher Education in Comparative Perspective*. London: Palgrave Macmillan.

Cashore, Benjamin, and Michael Howlett. 2007. Punctuating Which Equilibrium? Understanding Thermostatic Policy Dynamics in Pacific Northwest Forestry. *American Journal of Political Science* 51 (3): 532–551.

Chandler, J.A. 2014. France. In *Comparative Public administration*. London: Routledge.

Chiang, Jong-Tsong. 1991. From 'Mission-Oriented' to 'Diffusion-Oriented' Paradigm: The New Trend of U.S. Industrial Technology Policy. *Technovation* 11 (6): 339–356.

Chmiliar, L. 2010. "Multiple Case Design." in Mills, A. J., Durepos, G., Wiebe, E. (Eds.). Encyclopedia *of Case Study Research*. Thousand Oaks: SAGE Publications.

Choi, Sang Ok, Sang-Seok Bae, Sung-Wook Kwon, and Richard Feiock. 2008. County Limits: Policy Types and Expenditure Priorities. *The American Review of Public Administration* 40 (1): 29–45.

Chou, Meng-Hsuan, Jens Jungblut, Pauline Ravinet, and Martina Vukasovic. 2017. Higher Education Governance and Policy: An Introduction to Multi-Issue, Multi-Level and Multi-Actor Dynamics. *Policy and Society* 36 (1): 1–15.

Chung, Chao-chen. 2013. Government, Policy-Making and the Development of Innovation System: The Cases of Taiwanese Pharmaceutical Biotechnology Policies (2000–2008). *Research Policy* 42 (5): 1053–1071.

Cobis, Fabrizio. 2008. I Nuovi Processi Di Innovazione e l'azione Del Ministero Dell'Università e Della Ricerca. In *I Nuovi Processi Di Innovazione e l'azione Del Ministero Dell'Università e Della Ricerca*, 1000–1005.

Coccia, Mario, and Secondo Rolfo. 2010. New Entrepreneurial Behaviour of Public Research Organisations: Opportunities and Threats of Technological Services Supply. *International Journal Services Technology and Management* 13 (1/2): 134–151.

Considine, Mark, and Kamran Ali Afzal. 2010. Legitimacy. In *The SAGE Handbook of Governance*, ed. M. Bevir. Berkeley: University of California, SAGE Publications.

Corradi, Fiammetta. 2009. *Alla Ricerca Dell'eccellenza: Le Politiche per l'eccellenza Nell'istruzione Superiore in Quattro Paesi Europei*. LED, Edizioni universitarie di lettere, economia, diritto.

Curley, Cali, Richard Feiock, and Xu. Kewei. 2020. Policy Analysis of Instrument Design: How Policy Design Affects Policy Constituency. *Journal of Comparative Policy Analysis: Research and Practice* 22 (6): 536–557.

Cytermann, Jean-Richard. 2006. L'architecture de la loi organique relative aux lois de finances (LOLF) dans les domaines de l'éducation et de la recherche: choix politiques ou choix techniques? *Revue française d'administration publique* 117 (1): 85–93.

Dente, B. 2014. *Understanding Policy Decisions*. Cham: Springer.

Dermont, Clau, Karin Ingold, Lorenz Kammermann, and Isabelle Stadelmann-Steffen. 2017. Bringing the Policy Making Perspective in: A Political Science Approach to Social Acceptance. *Energy Policy* 108: 359–368.

Doern, G., and R. Phidd. 1983. *Canadian Public Policy: Ideas, Structures, Process*. Toronto: Methuen.

Donina, Davide, Michele Meoli, and Stefano Paleari. 2015. Higher Education Reform in Italy: Tightening Regulation Instead of Steering at a Distance. *Higher Education Policy* 28 (2): 215–234.

Dosso, Mafini. 2014. "Restructuring in France's Innovation System: From the Mission-Oriented Model to a Systemic Approach of Innovation." *LEM Papers Series.*

Dür, Andreas. 2008. Measuring Interest Group Influence in the EU: A Note on Methodology. *European Union Politics* 9 (4): 559–576.

Edler, Jakob, Paul Cunningham, Kieron Flanagan, and Philippe Laredo. 2013. *Innovation Policy Mix and Instrument Interaction: A Review*, Nesta Working Paper No. 13/20. London: Nesta.

Edler, Jakob, Paul Cunningham, A. Gök, and Philip Shapira. 2016. *Handbook of Innovation Policy Impact.* Cheltenham: Edward Elgar Publishing.

Edler, Jakob, and Jan Fagerberg. 2017. Innovation Policy: What, Why, and How. *Oxford Review of Economic Policy* 33 (1): 2–23.

Edler, Jakob, and Stefan Kuhlmann. 2008. Coordination within Fragmentation: Governance in Knowledge Policy in the German Federal System. *Science and Public Policy* 35 (4): 265–276.

Edquist, Charles. 2001. The Systems of Innovation Approach and Innovation Policy: An Account of the State of the Art. DRUID conference, Aalborg.

Edquist, Charles, and Susana Borrás. 2013. The Choice of Innovation Policy Instruments. *Technological Forecasting and Social Change* 80: 1513–1522.

Eparvier, Patrick; Turcat, Nicolas; Schoen, Antonie; Carat, Gerard; Nill, Jan. 2008. ERAWATCH Country Report 2008. An Assessment of Research System and Policies France

Eparvier, Patrick, Flora Giarracca, and Léonor Rivoire. 2009. *ERAWATCH Country Report 2009: France.*

Eparvier, Patrick, Olivier Mallet, and Léonor Rivoire. 2011. *ERAWATCH Country Reports 2011: France.*

Ergas, H. 1987. The Importance of Technology Policy. In *Economic Policy and Technological Performance*, ed. P. Dasgupta and P. Stoneman. Cambridge: Cambridge University Press.

European Commission. 1999. European Trend Chart on Innovation Country Report: Italy.

———. 2000a. European Trend Chart on Innovation - Country Report France.

———. 2000b. European Trend Chart on Innovation Country Report: Italy.

———. 2001. European Trend Chart on Innovation Country Report: Italy.

———. 2002a. European Trend Chart on Innovation Country Report: Italy.

———. 2002b. EUROPEAN TREND CHART Theme-Specific Country Report: France.

———. 2003. European Trend Chart on Innovation Country Report: Italy.

———. 2004a. European Trend Chart on Innovation Annual Innovation Policy Trends and Appraisal Report: France.

———. 2004b. European Trend Chart on Innovation Country Report: Italy.

———. 2005a. European Trend Chart on Innovation - Annual Innovation Policy Trends and Appraisal Report: France.

———. 2005b. European Trend Chart on Innovation Country Report: Italy.

———. 2006. European Trend Chart on Innovation Country Report: Italy.

Evangelista, Rinaldo. 2007. Rilevanza e Impatto Delle Politiche Dell'innovazione in Italia. I Risultati Delle Indagini CIS. *Econ. Politica Ind.* 1: 103–124.

Fagerberg, Jan. 2017. Innovation Policy: Rationales, Lessons and Challenges. *Journal of Economic Surveys* 31 (2): 497–512.

Flanagan, Kieron, Elvira Uyarra, and Manuel Laranja. 2011. Reconceptualising the 'Policy Mix' for Innovation. *Research Policy* 40 (5): 702–713.

Foray, D., D.C. Mowery, and R.R. Nelson. 2012. Public R&D and Social Challenges: What Lessons from Mission R&D Programs? *Research Policy* 41 (10): 1697–1702.

Foray, Dominique. 2009. *The New Economics of Technology Policy. Cheltenham*: Edward Elgar. London.

Fraussen, Bert. 2014. The Visible Hand of the State: On the Organizational Development of Interest Groups. *Public Administration* 92 (2): 406–421.

Freeman, Gary P. 2006. National Models, Policy Types, and the Politics of Immigration in Liberal Democracies. *West European Politics* 29 (2): 227–247.

French Court of Auditors. 2013. Le Financement Public de La Recherche, Un Enjeu National.

Gallo, Riccardo. 2016. Torniamo a Industriarci: A Novant'anni Dalla 'Grande Crisi' Guida.

Gallo, Riccardo, Mauro Mallone, and Vincenzo Zezza. 2008. Technology Transfer: The RIDITT Experience. *L'industria* 1: 207–226.

Gallo, Riccardo, and Francesco Silva. 2006. Le Condizioni per Crescere: Diagnosi e Proposte per Il Sistema Produttivo. Il Sole 24 ore.

Gassler, Helmut, Wolfgang Polt, and Christian Rammer. 2008. Priority Setting in Technology Policy: Historical Developments and Recent Trends. In *Innovation Policy in Europe: Measurement and Strategy*, ed. C. Nauwelaers and R. Wintjes. Cheltenham: Edward Elgar.

Giest, Sarah. 2016. The Challenges of Enhancing Collaboration in Life Science Clusters: Lessons from Chicago, Copenhagen and Singapore: Table 1. *Science and Public Policy* 44 (2): scw046.

Goetz, Klaus H. 2008. Governance as a Path to Government. *West European Politics* 31 (1–2): 258–279.

Griessen, Thomas, and Dietmar Braun. 2008. The Political Coordination of Knowledge and Innovation Policies in Switzerland. *Science and Public Policy* 35 (May): 277–288.

Gross, Catherine. 2007. Community Perspectives of Wind Energy in Australia: The Application of a Justice and Community Fairness Framework to Increase Social Acceptance. *Energy Policy* 35 (5): 2727–2736.

Hall, Peter A. 1993. Policy Paradigms, Social Learning, and the State: The Case of Economic Policymaking in Britain. *Comparative Politics* 25 (3): 275–296.

———. 2010. Historical Institutionalism in Rationalist and Sociological Perspective. In *Explaining Institutional Change: Ambiguity, Agency, and Power*, ed. J. Mahoney and K. Thelen, 204–224. Cambridge: Cambridge University Press.

Hartmann, Eva, and Poul F. Kjaer. 2015. *The Evolution of Intermediary Institutions in Europe*. London: Palgrave Macmillan UK.

Helmke, Gretchen, and Steven Levitsky. 2004. Informal Institutions and Comparative Politics: A Research Agenda. *Perspectives on Politics* 2 (4): 725–740.

Héritier, Adrienne, and Dirk Lehmkuhl. 2011. New Modes of Governance and Democratic Accountability. *Government and Opposition* 46 (1): 126–144.

Hofmann, Eva, Katharina Gangl, Erich Kirchler, and Jennifer Stark. 2014. Enhancing Tax Compliance through Coercive and Legitimate Power of Tax Authorities by Concurrently Diminishing or Facilitating Trust in Tax Authorities. *Law and Policy* 36 (3): 290–313.

Hood, Christopher. 1986. *The Tools of Government / Christopher C. Hood*. Chatham: Chatham House.

Hood, Christopher, and Helen Magretts. 2007. *The Tools of Government in the Digital Age*. Macmillan. Red Globe Press.

Howlett, M., and I. Mukherjee. 2018. *Routledge Handbook of Policy Design*. 1st ed. London: Routledge.

Howlett, M., M. Ramesh, and A. Perl. 2009. *Studying Public Policy: Policy Cycles & Policy Subsystems*. Oxford: Oxford University Press.

Howlett, Michael. 2000. Managing the 'Hollow State': Procedural Policy Instruments and Modern Governance. *Canadian Public Administration/ Administration Publique Du Canada* 43 (4): 412–431.

———. 2009. Governance Modes, Policy Regimes and Operational Plans: A Multi-Level Nested Model of Policy Instrument Choice and Policy Design. *Policy Sciences* 42 (1): 73–89.

———. 2018. Aligning Policy Tools and Their Targets: Nudging and Utility Maximization in Policy Design. In *Routledge Handbook of Policy Design*, ed. M. Howlett and I. Mukherjee, 1st ed., 106–121. *London*: Routledge.

Howlett, Michael, and Ishani Mukherjee. 2014. Policy Design and Non-Design: Towards a Spectrum of Policy Formulation Types. *Politics and Governance* 2 (2): 57.

Howlett, Michael, and M. Ramesh. 1993. Patterns of Policy Instrument Choice: Policy Styles, Policy Learning and the Privatization Experience. *Review of Policy Research* 12 (1–2): 3–24.

Howlett, Michael, M. Ramesh, and Giliberto Capano. 2020. Policy-Makers, Policy-Takers and Policy Tools: Dealing with Behaviourial Issues in Policy Design. *Journal of Comparative Policy Analysis: Research and Practice* 22 (6): 487–497.

Howlett, Michael, and Jeremy Rayner. 2013. Patching vs Packaging in Policy Formulation: Assessing Policy Portfolio Design. *Politics and Governance* 1 (2): 170.

IGAENR. 2012. *Etude Des Mécanismes d'allocation Des Moyens Humains et Financiers Aux Unités de Recherche Par Les Organismes de Recherche*. Direction de l'information légale et administrative.

Ingold, Karin, Isabelle Stadelmann-Steffen, and Lorenz Kammermann. 2018. The Acceptance of Instruments in Instrument Mix Situations: Citizens' Perspective on Swiss Energy Transition. *Research Policy* 48: 103694.

John, Peter, ed. 2012. *Analysing Public Policy*. 2nd ed. London: Routledge.

Jungblut, Jens. 2015. Bringing Political Parties into the Picture: A Two-Dimensional Analytical Framework for Higher Education Policy. *Higher Education* 69 (5): 867–882.

Kammermann, Lorenz, and Karin Ingold. 2019. Going beyond Technocratic and Democratic Principles: Stakeholder Acceptance of Instruments in Swiss Energy Policy. *Policy Sciences* 52 (1): 43–65.

Kassim, Hussein, and Patrick Le Galès. 2010. Exploring Governance in a Multi-Level Polity: A Policy Instruments Approach. *West European Politics* 33 (1): 1–21.

Knill, Christoph. 2001. *The Europeanisation of National Administrations: Patterns of Institutional Change and Persistence*. Cambridge: Cambridge University Press.

Koch, Christian. 2008. The Superministry Approach: Integrated Governance of Science, Technology and Innovation with Contracted Autonomy. *Science and Public Policy* 35 (4): 253–264.

Lai, Luigi. 2016. Innovation Policies and SMEs Development. In *Entrepreneurship: Antecedents and Effects "Przedsiębiorczość Międzynarodowa"*, ed. M. Kosała, M. Urbaniec, and A. Żur, vol. 2, no. 2, 147–163. Kraków: Cracow University of Economics.

Lanza, Orazio, and Kostas Lavdas. 2000. The Disentanglement of Interest Politics: Business Associability, the Parties and Policy in Italy and Greece. *European Journal of Political Research* 37 (2): 203–235.

Laranja, Manuel, Elvira Uyarra, and Kieron Flanagan. 2008. Policies for Science, Technology and Innovation: Translating Rationales into Regional Policies in a Multi-Level Setting. *Research Policy* 37 (5): 823–835.

Larrue, Philippe, Patrick Eparvier, and Sophie Bussillet. 2006. *Etude de l'impact Du Crédit Impôt Recherche*.

Lascoumes, Pierre, and Patrick Le Gales. 2007. Introduction: Understanding Public Policy through Its Instruments? From the Nature of Instruments to the Sociology of Public Policy Instrumentation. *Governance* 20 (1): 1–21.

Laver, Michael, and Kenneth Shepsle. 1996. The Social Context of Government Formation. In *Making and Breaking Governments. Cabinets and legislatures in Parliamentary Democracies*, ed. K. Laver and Shepsle Michael, 18–42. Cambridge: Cambridge University Press.

Lepori, Benedetto. 2011. Coordination Modes in Public Funding Systems. *Research Policy* 40 (3): 355–367.

Lepori, Benedetto, Emanuela Reale, and Emilia Primieri. 2017. *Public Funding Country Profile France Annex 11.*

Lindblom, C.E. 1959. The Science of 'Muddling Through.'. *Public Administration Review* 19 (2): 79–88.

Linder, Stephen H., and B. Guy Peters. 1989. Instruments of Government: Perceptions and Contexts. *Journal of Public Policy* 9 (01): 35.

Lissoni, Francesco, Peter Lotz, Jens Schovsbo, and Adele Treccani. 2009. Academic Patenting and the Professor's Privilege: Evidence on Denmark from the KEINS Database. *Science and Public Policy* 36 (8): 595–607.

Lissoni, Francesco, and Fabio Montobbio. 2015. The Ownership of Academic Patents and Their Impact. Evidence from Five European Countries. *Revue économique* 66 (1): 143–171.

Lowi, T.J. 1964. American Business, Public Policy, Case Studies, and Political Theory. *Word Politics* 16: 677–715.

Lucchese, Matteo, Leopoldo Nascia, Mario Pianta, and Mario Pianta. 2016. Industrial Policy and Technology in Italy. *Economia e Politica Industriale* 43 (3): 233–260.

Macdonald, Douglas. 2001. Coerciveness and the Selection of Environmental Policy Instruments. *Canadian Public Administration/Administration Publique Du Canada* 44 (2): 161–187.

De Maio, Adriano. 2011. *L' Innovazione Vincente*. Brioschi.

Di Maio, Michele. 2014. The Italian Economy, the Economic Crisis and Industrial Policy. In *Structural change, competitiveness and industrial policy: painful lessons from the European periphery*, ed. A. Teixeira and E. Silva. London: Routledge.

Majone, Giandomenico. 1989. *Evidence, Argument, and Persuasion in the Policy Process*. New Heaven: Yale University Press.

———. 1976. Choice among Policy Instruments for Pollution Control. *Policy Analysis* 2: 589–613.

Malerba, Franco. 1993. The National System of Innovation: Italy. In *National Innovation Systems: A Comparative Analysis.*

March, James G., and Johan P. Olsen. 1997. *Democratic Governance*. New York: Free Press.

———. 2004. *The Logic of Appropriateness*. Oslo: Arena.

Mathison, Sandra. 2005. Cross-Case Analysis. In *Encyclopedia or Evaluation*, ed. S. Mathison. Thousand Oaks: SAGE Publications.

Merton, R.K. 1938. Social Structure and Anomie. *American Sociological Review* 3 (5): 672–682.
Van der Meulen, Barend. 1998. Science Policies as Principal–Agent Games: Institutionalization and Path Dependency in the Relation between Government and Science. *Research Policy* 27 (4): 397–414.
Miles, M.B., and A.M. Huberman. 1994. *Qualitative Data Analysis: An Expanded Sourcebook*. Thousand Oaks: Sage Publications, Inc.
Ministére de l'education nationale de l'enseignment supérieur et de la Recherche. 2014. *Développement et Impact Du Crédit d'impôt Recherche: 1983–2011*.
Ministero dell'Istruzione dell'Università e della Ricerca. 2005. *Programma Nazionale per La Ricerca*.
MISE. 2010. *Gli Incentivi Alle Attività Produttive. Effetti Della Legge 46/82 Sull'innovazione e Sull'industrializzazione Dei Prototipi Di Ricerca*.
Musselin, Christine (1958-........). 2017. *La Grande Course Des Universités*. edited by Sciences Po Presse.
Musselin, Christine, and Catherine Paradeise. 2009. France: From Incremental Transitions to Institutional Change. In *University Governance: Western European Comparative Perspectives*, ed. Catherine Paradeise, Emanuela Reale, Ivar Bleiklie, and Ewan Ferlie, 21–49. Cham: Springer Verlag.
Mustar, Philippe, and Philippe Laredo. 2002. Innovation and Research Policy in France (1980–2000) or the Disappearance of the Colbertist State. *Research Policy* 31 (1): 55–72.
Mustar, Philippe, and Philippe Larédo. 2002. Innovation and Research Policy in France (1980–2000) or the Disappearance of the Colbertist State. *Research Policy* 31 (1): 55–72.
Nascia, Leopoldo, and Mario Pianta. 2013. *ERAWATCH Country Reports 2013: Italy*.
Nascia, Leopoldo, Bianca Potì, and Emanuela Reale. 2012. *ERAWATCH Country Report 2012: Italy*.
Nauwelaers, Claire, and René Wintjes. 2008. *Innovation Policy in Europe: Measurement and Strategy*. Cheltenham: Edward Elgar.
Nelson, Richard R. 1993. *National Innovation Systems: A Comparative Analysis*. New York: Oxford University Press.
North, D.C. 1990. *Institutions, Institutional Change and Economic Performance*. Cambridge: Cambridge University Press.
OECD. 2005. *Governance of Innovation Systems*. OECD.
———. 2014. *Reviews of Innovation Policy: France 2014*. OECD.
Olsson, Jan, and Joachim Åström. 2003. Why Regionalism in Sweden? *Regional & Federal Studies* 13 (3): 66–89.
Palier, Bruno. 2007. Tracking the Evolution of a Single Instrument Can Reveal Profound Changes: The Case of Funded Pensions in France. *Governance* 20 (1): 85–107.

Pelkonen, Antti, Tuula Teräväinen, and Suvi-Tuuli Waltari. 2008. Assessing Policy Coordination Capacity: Higher Education, Science and Technology Policies in Finland. *Science and Public Policy* 35 (4): 241–252.
Perry, Beth, and Tim May. 2007. Governance, Science Policy and Regions: An Introduction. *Regional Studies* 41 (8): 1039–1050.
Peters, B G. 1998a. Managing Horizontal Government: The Politics of Coordination. Canadian Centre for Management Development (1998).
Peters, B. Guy. 1998b. *Comparative Politics*. London: Macmillan Education UK.
———. 2000. The Politics of Tool Choice. In *The Tools of Government: A Guide to the New Governance*, ed. L. Salamon, 1st ed. New York: Oxford University Press.
———. 2012. Governance As Political Theory. In *The Oxford Handbook of Governance*, ed. Levi-Faur. New York: Oxford University Press.
Pierantozzi, Valentina. 2007. *ERAWATCH Analytical Country Report 2007: Italy.*
Pochet, Philippe, and Caroline de la Porte. 2002. *Building Social Europe through the Open Method of Co-Ordination*. Brussels: P. Lang.
Potì, Bianca, and Emanuela Reale. 2006. *Project Funding. Italy.*
———. 2007. Changing Allocation Models for Public Research Funding: An Empirical Exploration Based on Project Funding Data. *Science and Public Policy* 34 (6): 417–430.
———. 2010. *Erawatch Country Reports 2010: Italy.*
———. 2011. *ERAWATCH Country Report 2011: Italy.*
Poti, Bianca, Emanuela Reale, and Valentina Pierantozzi. 2008. *ERAWATCH Country Report. An Assessment of Research System and Policies. Italy.*
Prange, Heiko. 2008. Explaining Varieties of Regional Innovation Policies in Europe. *European Urban and Regional Studies* 15 (1): 39–52.
Reale, Emanuela, and Lucio Morettini. 2017. *Public Funding Country Profile Italy Annex 11.*
Reale, Emanuela, and Bianca Potì. 2009. Italy: Local Policy Legacy and Moving to an 'In Between' Configuration. In *University governance: Western European comparative perspectives*, ed. Catherine Paradeise, Emanuela Reale, Ivar Bleiklie, and Ewan Ferlie, 77–102. Cham: Springer Verlag.
Reillon, V. 2015. *Overivew of EU Funds for Research and Innovation*. Brussels.
Rimkutė, Dovilė, and Markus Haverland. 2015. How Does the European Commission Use Scientific Expertise? Results from a Survey of Scientific Members of the Commission's Expert Committees. *Comparative European Politics* 13 (4): 430–449.
Ringeling, Arthur B. 2005. Instruments in Four: The Elements of Policy Design. In *Designing Government: From Instruments to Governance*, ed. P. Eliadis, M.M. Hills, and M. Howlett, 185–202. Kingston: O. McGill-Queens University Press.

Rist, Ray C. 1998. Choosing the Right Policy Instrument at the Right Time: The Contextual Challenges of Selection and Implementation. In *In Carrots, Sticks and Sermons: Policy Instruments and Their Evaluation*, ed. M.-L. Bemelmans-Videc, R. Rist, and E. Vedung. New Brunswick: Transaction.

Ritchie, J., and J. Lewis. 2003. *Qualitative Research Practice: A Guide for Social Science Students and Researchers*. Thousand Oaks: SAGE Publications.

Rogge, Karoline S. 2018. Designing Complex Policy Mixes. In *Routledge Handbook of Policy Design*, ed. M. Howlett and I. Mukherjee, 1st ed., 34–58. London: Routledge.

Rogge, Karoline S., and Kristin Reichardt. 2016. Policy Mixes for Sustainability Transitions: An Extended Concept and Framework for Analysis. *Research Policy* 45 (8): 1620–1635.

Rueschemeyer, Dietrich, and John D. Stephens. 1997. Comparing Historical Sequences: A Powerful Tool for Causal Analysis. *Comparative Social Research* 16: 55–72.

Sabatier, Paul A. 1986. Top-Down and Bottom-Up Approaches to Implementation Research: A Critical Analysis and Suggested Synthesis. *Journal of Public Policy* 6 (1): 21–48.

Salamon, Lester. 2002. *The Tools of Government: A Guide to the New Governance*. Oxford University Press.

———. 2011. The New Governance and the Tools of Public Action. 28(5). Symposium - Redefining the Public Sector: Accountability and Democracy in the Era of Privatization.

Sanz-Mendez, Luis, and Susana Borrás. 2001. Explaining Changes and Continuity in EU Technology Policy: The Politics of Ideas. In *Changing European Research System*, ed. Imon Dresner and Nigel Gilbert. Aldershot: Ashgate Press.

Schmidt, Tobias S., and Sebastian Sewerin. 2019. Measuring the Temporal Dynamics of Policy Mixes – An Empirical Analysis of Renewable Energy Policy Mixes' Balance and Design Features in Nine Countries. *Research Policy* 48 (10): 103557.

Schneider, Anne, and Helen Ingram. 1990. Behavioral Assumptions of Policy Tools. *The Journal of Politics* 52 (2): 510–529.

———. 1993. Social Construction of Target Populations: Implications for Politics and Policy. *American Political Science Review* 87 (2): 334–347.

Schneider, Anne, and Mara Sidney. 2009. What Is Next for Policy Design and Social Construction Theory? *Policy Studies Journal* 37 (1): 103–119.

Schoen, Antonie, Gerard Carat, and Nill Jan. 2008. *ERAWATCH Analytical Country Report 2007: France*.

Schreier, Margrit. 2012. *Qualitative Content Analysis in Practice*. London: SAGE Publications Ltd.

Seawright, Jason, and John Gerring. 2008. Case Selection Techniques in Case Study Research: A Menu of Qualitative and Quantitative Options. *Political Research Quarterly* 61 (2): 294–308.

Seravalli, G. 2009. *Competitive European Regions through Research and Innovation. Different Theoretical Approaches to Innovation Policies.*

Silva, Francesco. 2007. La Politica Industriale Del Governo Prodi. *Economia e Politica Industriale* 2: 7–19.

Simons, Arno, and Jan-Peter Voß. 2018. The Concept of Instrument Constituencies: Accounting for Dynamics and Practices of Knowing Governance. *Policy and Society* 37 (1): 14–35.

Sirilli, Giorgio. 2010. *La Produzione e La Diffusione Della Conoscenza. Ricerca, Innovazione e Risorse Umane.*

Smits, Ruud, Philip Shapira, and Stefan Kuhlmann. 2010. *The Theory and Practice of Innovation Policy: An International Research Handbook.* Cheltenham: Edward Elgar.

Spence, R.E. 2014. Italy. In *Comparative public administration*, ed. J.A. Chandler, 2nd ed. London: Routledge, Taylor & Francis Group.

Stoppino, Mario. 2001. *Potere e Teoria Politica.* Giuffre.

Tamtik, Merli. 2016. Institutional Change Through Policy Learning: The Case of the European Commission and Research Policy. *Review of Policy Research* 33 (1): 5–21.

Thèves, Jean, Benedetto Lepori, and Philippe Larédo. 2007. Changing Patterns of Public Research Funding in France. *Science and Public Policy* 34 (6): 389–399.

Tödtling, Franz, and Michaela Trippl. 2005. One Size Fits All? Towards a Differentiated Regional Innovation Policy Approach. *Research Policy* 34: 1203–1219.

———. 2012. Transformation of Regional Innovation Systems: From Old Legacies to New Development Paths. 52nd Congress of the European Regional Science Association: "Regions in Motion - Breaking the Path", 21–25 August 2012, Bratislava, Slovakia.

Tosun, Jale. 2018. Investigating Ministry Names for Comparative Policy Analysis: Lessons from Energy Governance. *Journal of Comparative Policy Analysis: Research and Practice* 20 (3): 324–335.

Traù, Fabrizio. 2009. Logica Di Un Sistema Di Sostegno Alle Imprese. La Politica Industriale Italiana Dal Secondo Dopoguerra a 'Industria 2015'. *ECONOMIA E POLITICA INDUSTRIALE* 4.

Trebilcock, Michael J., and Douglas G. Hartle. 1982. The Choice of Governing Instrument. *International Review of Law and Economics* 2 (1): 29–46.

Trein, Philipp. 2017. A New Way to Compare Horizontal Connections of Policy Sectors: 'Coupling' of Actors, Institutions and Policies. *Journal of Comparative Policy Analysis: Research and Practice* 19 (5): 419–434.

Varone, Frédéric, and Bernard Aebischer. 2001. Energy Efficiency: The Challenges of Policy Design. *Energy Policy* 29 (8): 615–629.

Vedung, Evert. 1998. Policy Instruments: Typologies and Theories. In *Carrots, Sticks and Sermons: Policy Instruments and Their Evaluation*, ed. E. Vedung,

M.L. Bemelmans-Videc, and R.C. Rist, 21–58. New Brunswick: Transaction Publishers.

Van de Ven, A.H., and R. Garud. 1989. A Framework for Understanding the Emergence of New Industries. In *Research on Technological Innovation and Management Policy*, ed. R. Rosenbloom and R. Burgelman. Greenwich: JAI Press.

Verhoest, Koen, and Geert Bouckaert. 2005. "Machinery of Government and Policy Capacity." in Painter, M., Pierre, J *(2005)* Challenges to State Policy Capacity, edited by Basingstoke: Palgrave.

Voß, Jan-Peter, and Arno Simons. 2014. Instrument Constituencies and the Supply Side of Policy Innovation: The Social Life of Emissions Trading. *Environmental Politics* 23 (5): 735–754.

———. 2018. Instrument Constituencies: Promoting Policy Designs. In *Routledge Handbook of Policy Design*, ed. M. Howlett and I. Mukherjee, 1st ed., 180–200. London: Routledge.

van Waarden, F. 1995. Persistence of National Policy Styles: A Study of Their Institutional Foundations. In *Convergence or Diversity*, ed. B. Unger and F. van Waarden, 333–372. Vermont: Ashgate Publishing Company.

Wilson, Carter. 2000. Policy Regimes and Policy Change. *Journal of Public Policy* 20 (3): 247–274.

Wilson, James. 1973. *Political Organizations.* New York: Basic Book.

———. 1980. *The Politics of Regulation.* New York: Basic Book.

Woodside, Kenneth. 1983. The Political Economy of Policy Instruments: Tax Expenditures and Subsidies. In *The Politics of Canadian Public Policy*, ed. Michael M. Atkinson and Marsha A. Chandler, 173–197. Toronto: University of Toronto Press.

———. 1986. Policy Instruments and the Study of Public Policy. *Canadian Journal of Political Science / Revue Canadienne de Science Politique* 19 (4): 775–793.

Zaparucha, Elisabeth. 2010. *ERAWATCH Country Report 2010: France.*

Bibliography

Borrás, Susana. 2009. The Widening and Deepening of Innovation Policy: What Conditions Provide for Effective Governance? *Papers in Innovation Studies 2009/2, Lund University, CIRCLE - Center for Innovation, Research and Competences in the Learning Economy.*

Buchs, Milena. 2007. New Governance in European Social Policy: The Open Method of *Coordination*. Palgrave Macmillan.

Capano, G., JE. Rayner, and A. R. Zito. 2012. Governance from the Bottom up: Complexity and Divergence in Comparative Perspective. *Public Administration* 90(1):56–73.

Chung, Chao-chen. 2013. Government, Policy-Making and the Development of Innovation System: The Cases of Taiwanese Pharmaceutical Biotechnology Policies (2000–2008). *Research Policy* 42(5):1053–71.

Flanagan, Kieron, Elvira Uyarra, and Manuel Laranja. 2011. Reconceptualising the 'Policy Mix' for Innovation. *Research Policy* 40(5):702–13.

Héritier, Adrienne, and Dirk Lehmkuhl. 2011. New Modes of Governance and Democratic Accountability. *Government and Opposition* 46(1):126–44.

Olsson, Jan, and Joachim Åström. 2003. Why Regionalism in Sweden? *Regional & Federal Studies* 13(3):66–89.

Perry, Beth, and Tim May. 2007. Governance, Science Policy and Regions: An Introduction. *Regional Studies* 41(8):1039–50.

Pochet, Philippe, and Caroline de la Porte. 2002. *Building Social Europe through the Open Method of Co-Ordination*. Brussels: P. Lang.

Prange, Heiko. 2008. Explaining Varieties of Regional Innovation Policies in Europe. *European Urban and Regional Studies* 15(1):39–52.

Rellion, Vincent. 2015. *Overview of EU Funds for Research and Innovation*. Brussels.

Seravalli, Gilberto. 2009. *Competitive European Regions through Research and Innovation. Different Theoretical Approaches to Innovation Policies.*

Tamtik, Merli. 2016. Institutional Change Through Policy Learning: The Case of the European Commission and Research Policy. *Review of Policy Research* 33(1):5–21.

Thèves, Jean, Benedetto Lepori, and Philippe Larédo. 2007. Changing Patterns of Public Research Funding in France. *Science and Public Policy* 34(6):389–99.

Tödtling, Franz, and Michaela Trippl. 2005. One Size Fits All? Towards a Differentiated Regional Innovation Policy Approach. *Research Policy* 34:1203–19. 52nd Congress of the European Regional Science Association: "Regions in Motion - Breaking the Path", 21–25 August 2012, Bratislava, Slovakia.

Tödtling, Franz, and Michaela Trippl. 2012. *Transformation of Regional Innovation Systems: From Old Legacies to New Development Paths.*

CHAPTER 4

Research and Innovation Policy Design in France

French R&I policies have been historically characterized by a tradition of highly centralized public R&D programs. These strategies were organized through the coordination between large public research organizations and national enterprises specialized in sectors like aerospace, nuclear energy, and telecommunications. The system has undergone consistent changes in the last decades, due to an acquired relevance of the structures devoted to the support of technology transfer. With the establishment of a National Research Agency (2005), public research moved toward more competitive-oriented funding systems, which was then combined with the increased autonomy of universities (2007 reforms). Despite different structural reforms in R&I, the sector remains affected by an intricate institutional landscape (Dosso 2014). Therefore, despite increasing attention toward technology transfer and the creation of specific organizations for its development, the system can still be characterized as mainly mission-oriented (Berger 2016).

The following paragraphs illustrate the evolution of French R&I governance from the late 1990s until 2013, by decomposing the different combinations of actor interactions, considering the evolution of national policy mix. As described in Fig. 4.1, this chapter illustrates the process that culminated with the mandate given to Henri Guillaume for drafting *Rapport Guillaume*, which triggered various reforms in R&I such as the Law on Innovation and Research (1999) and the Innovation Plan (2003). Then,

C. Acciai, *Policy Design for Research and Innovation*, International Series on Public Policy,
https://doi.org/10.1007/978-3-031-36628-4_4

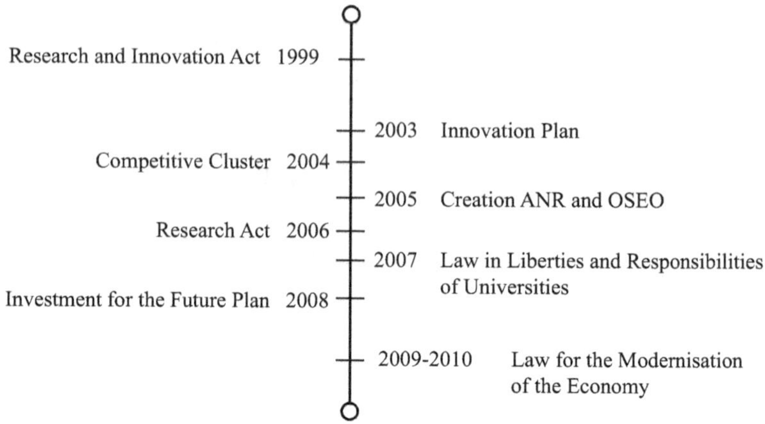

Fig. 4.1 The chronology of R&I policies (France)

the chapter will discuss the fragmented landscape of measures supporting SMEs, the design of the Competitive cluster strategy (2004), and the parallel developments in the Research sector, with the adoption of the 2006 Research Act and the creation of different public agencies. Finally, it will touch upon the 2007 Law on the Liberties and Responsibilities of Universities and the Law for the Modernization of the Economy, concluding with a discussion on the Investment for the Future plan.

As discussed in Chap. 1, in this research we are adopting a broad perspective on the study of R&I policies because we believe that both knowledge production and knowledge application systems should be considered jointly. Nevertheless, the interactions between these two spheres might not be easy to build. Especially for France, which has been historically characterized by a separation between "research"—considered as a privilege of the public realm—and "innovation," perceived as an activity belonging to the private sphere (European Commission 2004). Therefore, in Fig. 4.1 and in the structure of this chapter, we decided to differentiate between those R&I strategies that attempted to adopt an encompassing, holistic-oriented approach (Borras, Edquist, 2019), from those which tend to be focused on either the research or the innovation component. The former refers to an R&I policy mix targeting different R&I performers, with an emphasis on creating coordination between these two subsystems. The latter refer to those strategies adopting a siloed, or segmented, approach in organizing "research" and "innovation" interventions. To fully comprehend the features of the different R&I instrument mixes implemented across different strategies, we invite the reader to look at the supplementary material in the appendix of this chapter. Here, we decompose each specific R&I intervention according to its instrument mix components, which are then described and classified according to the proposed R&I instrument typology and the target to whom they are addressed.

4.1 The National Governance of R&I

The governance of French R&I policies hinges upon two different policymaking institutions: the Ministry of Higher Education and Research and the Ministry of Economy, Finance, and Industry. The great difficulty embedded in these ministerial layouts is the required coordination between these two areas of responsibility (OECD 2014). At the highest level of the R&I governance sits the President of the Republic and the prime minister, usually advised by the High Council for Research and Technology (HCST), made up of established personalities in the research and innovation sector. The minister of higher education and research, together with the minister of economy, finance, and industry, share the primary positions in national R&I policymaking tasks. Indeed, with the contribution of sectorial ministries, they all draft the National Strategy for Research and Innovation, which then defines the funding streams toward different operators. After the implementation of the Investment for the Future Plan (2008), a new organization found its place in the governance of the sector: the General Commissariat for Investment (CGI). This committee, working in close connection with the prime minister, formulated the thematic allocation of competitive funding.

The main R&I national operators are the National Research Agency (ANR), created in 2005 with the task of funding research primarily on a competitive basis, which also became the manager of the Investment for the Future program; BPI France, a public investment bank providing both funding (e.g., loans and guarantees) and consultancy services for the development of innovation projects in the private sector; and the *Caisse des Dépôts et Consignations* (CDC Deposits and Consignments Fund), a state bank funding company and heavily involved in innovation and SMEs financing (OECD 2014). Then, among R&I performers, there are public research organizations and higher education institutions, together with Research and Higher Education Clusters (which basically group these actors on a thematic, and geographical, principle). Finally, hybrid organizations of R&I performers including (potentially) private and public operators, like competitive clusters, Technological Transfer Acceleration Companies, and, finally, the different private R&I performers.

The Ministry of Higher Education and Research (MESR) and the Ministry with delegated competences for Industry represent the two fundamental institutions of the national governance of R&I. This ministerial organizational layout suggests how French R&I policies tend to be inherently made up of two substantive components (research and technology development), which can be differently integrated according to the capacity of different cabinets.

At the time of approval of the 1982 Law on Research, which introduced a financial division between public research organizations and their parent (functional) ministers, the competences for research, economy, finance, and industrial policies were merged within a unique ministerial organizational structure. However, in parallel with the gradual dismissal of the State's role in national Grand Programs, the two ministries have been separated. The Ministry of Industry has been merged within the broader Ministry of Economy and Finances (MINEFI), and industrial policies have since become less and less a matter of direct, targeted public intervention (Mustar and Larédo 2002).

4.2 Encompassing R&I Strategies

At the end of the 1990s, the French government undertook an extensive debate about the reorganization of its national R&I system, whose last systemic reform dated back to the 1982 *Loi de programmation et orientation de la recherche*.[1] The discussion was triggered on the one side by experts' reflections on the need to start evaluating university performance (Rapport Croizer, 1990). While on the other side, the association of national university chancellors stressed the importance of reforming the organization of their funding system (*Conférence des présidents d'université*, 1994).[2] In that same period of time, France was approaching the end of a historical transition phase which set off the end of major state-led public research (and industrial) programs.

[1] Loi n° 82–610 du 15 juillet 1982, d'orientation et de programmation pour la recherche.

[2] This text called for a "overhaul" of the university financing system and greater participation of local authorities and business actors (https://www.vie-publique.fr/politiques-publiques/enseignement-superieur-universite/chronologie/).

The national R&I system was welcoming the new century with many fallacies, like a low efficiency of technology transfer structures, low mobility of researchers, inadequacy of legal, financial, and fiscal conditions for the creation of innovative companies (European Commission 2002). Consequently, a mandate of reforming the national R&I governance was jointly given by the Ministry of Research, the Ministry of Economy, Finance, and Industry, and the Secretary of State for the Industry[3] to a national expert: Henry Guillaume.[4]

The resulting *Rapport Guillaume* (1998) highlighted some of the general difficulties French policymakers were facing in organizing and boosting interactions between public research and the industrial system, namely:

- a marked division between public research organizations and universities and engineering schools. These issues were closely related to the mixed nature of the national higher education and research system;
- the complexities embedded in the instrument for technology transfer, which often represented a barrier to the access of SMEs;
- the inadequacy of investments in risk-capitals (venture capitals and seed-capitals), still insufficient to cover the start-up phase of new technological enterprises;
- the lack of a consolidated national strategy for the coordination, implementation, evaluation, and support of industrial research;
- the lack of diversification in state investments, still focused on a limited number of industrial groups and sectors.[5]

To address the weaknesses highlighted by this report, policymakers organized a national consultation with R&I stakeholders, *Les Assises de l'Innovation* (1998). This was preceded by seven regional meetings, focusing on specific technological areas, organized by research organizations in liaison with local and regional actors. Consultations between policymakers and R&I stakeholders (policy takers) are a practice often adopted also in other historical phases of French R&I policymaking. These meetings are

[3] Respectively: Claude Allegre, Dominique Strauss-Khan, Christian Pierret.

[4] The honorary president of ANVAR and vice-president of ERAP, or *Entreprise de recherche et activités pétrolières*—enterprises for oil research and activities—a state-owned French petroleum company.

[5] Specifically, for many years, research priorities targeted raw materials, energy independence, and defense (European Commission 2002), as confirmed by Fig. 3.1.

meant to gather opinions and suggestions to enrich the design of national R&I strategies for involved policy takers.

4.2.1 The Law on Innovation and Research

The suggestions provided by the Guillaume report have been, (almost) directly, channeled into the 1999 Innovation Act (*Loi sur l'Innovation et la recherche*[6] also known as *Loi Allegre*).

This policy aimed at breaking down the deeply rooted mutual deference between academia and industry, while providing new tools for risk-capital development. Its instrument mix (Table 4.1, Appendix) hinged upon measures supporting the mobility of researchers toward private firms. The mix encouraged their participation in the establishment of innovative enterprises by easing their participation in the creation (and management) of innovative firms.[7]

By providing universities with the opportunity to create their own commercial and industrial business services (SAIC *Services d' activites industrielles et commerciales*), while developing projects with the industrial sector and regional partners (RRIT *Reseaux de Recherche et d'Innovation Technologique*, CNRT *Centres Nationaux de Recherche Technologique*, Regional Incubators), the mix aimed at enhancing the cooperation between research and industrial sectors. Finally, the reform introduced innovative solutions to back up venture capital development (co-investment funds for young enterprises and seed-capital development), to support private R&D expenses (CIR *Crédit Impôt Recherche* amendments), and to promote the creation of new technologically based firms (national competition for the creation of new technology-based firms). The instrument mix of the *Loi Allegre* is characterized by an exhortatory attitude. Its underlining logic goes in the direction of overcoming the divide between public research and the industrial world, as highlighted by the Guillaume Report. In line with that, the mix addresses all the different industrial and research stakeholders involved in the national R&I governance. By mostly using information and expenditure instruments, it

[6] Loi n. 99–587 du 12 juillet 1999 sur l'innovation et la recherche.

[7] Specifically, by extending the power of the existent legislation, the 1999 Law on Innovation and Research, in Article 2, increased the freedom of public research performers in the commercialization of their discoveries, amending the previous regulation provided by Loi n. 82–610 du 15 juillet 1982.

provides recipients with alternative strategies, and related funding opportunities, to establish collaborative networks of private and public innovation stakeholders. Finally, this was combined with an increased sensibility about providing a more favorable institutional and legal framework for the cooperation between innovation actors, as well as for the development of research results by public actors (especially in the form of start-up companies).

4.2.2 *The Innovation Plan*

The *Loi Allegre* has dominated the R&I policy agenda until 2003–2004. During the enactment of this strategy, the country underwent a period of intensive transformations in national R&I governance and instrumentation. The second stage of this systemic transformation was to combine national innovation strategies with economic and industrial interventions. In December 2002, the minister delegated for industry and the minister delegated for research and new technologies presented a plan aimed at strengthening the measures undertaken with the Innovation Act. Also, in this case, the strategy was discussed by R&I stakeholders in a national consultation which lasted three months. Meanwhile, the involved ministers went through a series of individual meetings with industrial managers, public and private researchers, experts, and investors. This process culminated with the *Plan pour l'innovation*, published in April 2003. The plan was made up of six axes of intervention: encouraging business-oriented investors (SUIP *Société Unipersonnelle d'Investissement Providentiel*), fiscal support for young R&D-intensive enterprises (JEI *Jeune Entreprises Innovantes*), and the reorganization of the system supporting research activities inside enterprises. This involved both the extension of already existent instruments (CIFRE[8] *convention industrielle de formation par la recherche*) and the organization of internal public competences in this sector, with new competences attributed to the ANVAR (National Agency for the Promotion of Research—*Agence nationale de valorisation de la recherche*). According to the Innovation Plan, the production of scientific and technological innovation must lay its foundations in the cooperation between public research and small private entities, like SMEs or start-ups. The instrument mix is a balanced combination between expenditure and fiscal instruments mainly addressed to R&I stakeholders from the indus-

[8] This instrument was put into force since 1981.

trial world. It is characterized by a low degree of coercion and a marked preference for both automatic and non-automatic delivery structures. Most importantly, this plan complements many of the instruments adopted with the previous *Loi Allegre*. Indeed, the latter aimed at encouraging R&I performers, providing a mix rich in instruments of the information type, while the Innovation Plan developed different economic and fiscal instruments directed at supporting the cooperation between researchers and the (small) industrial world. Such a design logic is combined with the evolution of the institutional monitoring system. Indeed, the ANVAR was first reformed into an agency with the specific task of ensuring coherence between national and local R&I measures. Then, a greater institutional reorganization took place (with the merger of BDPME—Development bank for SMEs—and ANVAR itself) with the creation of OSEO (January 2005) as the principal innovation public agency in charge of the implementation and monitoring of national policy, especially for enterprises (European Commission 2005).

4.2.3 *The Investment for the Future Plan*

Following the implementation of these two systemic reforms, various sector-specific measures have been implemented (Research Act, 2006; Law on the Liberties and Responsibilities of Universities, 2007: SME Pact, 2005; Law for the Modernization of the Economy, 2008). Policymakers got back to a holistic R&I policy design approach in 2007, with the introduction of the National Strategy for Research and Innovation (*Stratégie Nationale de Recherche et d'Innovation* SNRI). This is a five-year policy platform aimed at building a coordinated response to Great Societal Challenges. However, since it neither allocated a budget nor set operative goals (OECD 2014), it ended up resulting in a smooth exercise of voluntary, and somehow ceremonial, attempts to align the policy agenda of different ministries. Despite that, this (mild) programmatic effort was still influential in setting the stage for the development of a holistic R&I strategy under the Sarkozy presidency: the "*Investissement pour l'avenir*".

The Investment for the Future Plan (PIA) focused on improving the long-term growth potential of the French economy in the 2010–2020 period. It was inspired by the recommendations provided in a report drafted by a commission of experts chaired by Juppé and Rocard (two former prime ministers). More specifically, this committee was set up by the French government, following the 2008 economic and financial crisis, in order to identify the priority sectors for new strategic investment plans (Eparvier et al. 2009). The main recommendations of the expert report were focused on investing in higher education and research, while also prioritizing specific thematic areas for boosting national R&I systems. It was articulated across five sectors of priority: higher education and research, industries and SMEs, sustainable development, and the digital economy. With the implementation of this long-term innovation program, an innovative governance actor was also established: the *Commissariat Général à l'investissement* (CGI). This committee was responsible for steering and coordinating the overall coherence of the PIA scheme; its core tasks were focused on coordinating interministerial policy design under the authority of the prime minister as well as the cooperation among other governmental bodies responsible for the distribution of funds. It also supervised transparency and quality in the selection procedures, the allocation of funds to existent operators, as well as the overall coherence of the strategy.

The governance of the PIA strategy involved different intermediary actors: the ANR was responsible for the higher education and research sectors,[9] the Environment and Energy Management Agency (ADEME)[10] coordinated actions on energy and ecological transition, while OSEO (later BPI France) focused on supporting companies and industries. On top of this system of actors, the Commissariat-General (CGI) was overseeing the portfolio of national public investments across the different thematic areas. The plan was organized through different competitive calls for national projects around the thematic area managed by the CGI and its operators. The implementation was organized as follows: a first phase involved drafting agreements and defining the coordination strategies between the prime minister and the different intermediary agencies (ANR and OSEO/BPI France). The second phase focused on the organization

[9] The agency oversaw the management of calls for proposals, setting up the evaluation and selection process, preparing the grant agreements, funding the selected teams, monitoring the activities in their development, and providing the impact analysis.

[10] *Agence de l'environnement et de la maitrise de l'énergie*. It was an EPIC, which participated in the implementation of public policies in the fields of environment, energy, and sustainable development. The agency also supported project financing, from research to implementation, in the following areas: waste management, soil conservation, energy efficiency and renewable energies, raw materials, air quality, the fight against noise, the transition toward the circular economy, and the fight against food waste (https://www.ademe.fr/lademe. Last accessed May 2019).

and management of the calls for projects and the plan was structured across nine core strategies: financing of equipment for research/technological infrastructures (EQUIPEX); supporting leading labs, or public research institutions, aimed at competing for international funding (IDEX, IRT, LABEX, IEED); financing of research in health, biotechnology, and biomedical fields (IHU; Health and Biotechnology actions); and creating interface structures between academia and industry (SAAT; France Brevet). The overall instrument mix was characterized by a high share of expenditure-oriented and low coercive instruments. This instrumentation was combined with a high share of targeted funds aiming at supporting the increasing professionalization of the national bureaucracy with the prominent role of public agencies like ANR, OSEO, and ADEME. However, with the design of the PIA, the establishment of the commissariat with an intermediary role between operative agency and policymakers, the national R&I governance developed toward a new model in which intermediary agencies have the task of making policy instruments operational under the supervision of a "chief agency" (CGI). The delegation of both implementation and supervision policy tasks to intermediary agencies has contributed to a deeper evolution of the central state bureaucracy. Indeed, with the introduction of the PIA central administration, through its national public investment bank (CDC), this contributes to the development of innovative projects by acquiring equity participations in the capital of the SATT[11] (33% of their capital). This system can be defined as a "holding model" (Eparvier et al. 2011), where the central state administration, through the provision of different funds, possesses financial stakes in innovation organizations without the cost associated with their management. So, basically, the state has become an investor in innovation projects.[12]

[11] Technological Transfer Acceleration companies—*Société d'accélération du transfert technologique.*

[12] The Investment for the Future Plan has been maintained and extended in the following presidencies as well. The second wave of PIA (PIA 2), in 2014 under Hollande, slightly reoriented the strategy toward a more prominent role of the digital economy, and it increased the number and the specificity of the sectors of investments. The last wave (PIA 3) was implemented by Macron in 2017, and it became part of the broader *Grand Plan Pour L'investissement strategy*, with the goal of improving cooperative behaviors in R&D and related areas. Differently from the former strategy, the PIA 3 reduced the number of sector of investments and transformed the CGI in the *Secrétatiat General Pour l'Investissement* (SGPI).

4.3 The Reforms of Public Research and Higher Education Policies

The French public research system is a ***mixed research system*** because it is organized across both universities and public research organizations (the largest of which is the CNRS—*Centre national de la recherche scientifique* or French National Center for Scientific Research), and, unlike many other European countries, the latter play a greater role in structuring the public research scenario (OECD 2014). Historically, national public research organizations used to be heavily supported by the state and were focused on thematic missions, which happened to be closely related with core national areas of industrial specialization. Moreover, the same national industrial actors were the main entities responsible for the development and utilization of the research undertaken by the same public organizations (Mustar and Larédo 2002). Such a model of publicly steered research-to-innovation allowed for key technological breakthroughs in some sectors (e.g., space, railways). However, its limitations become clear with the emergence of new technological sectors, for example, ICTs (European Commission 2000). The Law on Innovation and Research (1999) and the Innovation Plan (2003) represented a first attempt to bridge the gaps of this model. Both reforms included an instrumentation aimed at raising innovation awareness among R&I stakeholders while supporting the creation of small innovative companies. However, their instruments mix only partially influenced the national public research sector, whose governance structure dated back to the 1982 Research Act. In an attempt to tackle these issues, policymakers drafted the *Pacte Pour la recherche* (2005), an action program aimed at re-organizing the French public research sector (Arnold 2007). The first proposal of the reform was outlined in the Innovation Plan presented by the Ministry delegated for Research and New Technologies, in 2003, and was expected to be launched in the second quarter of 2005. However, its first draft faced strong criticisms by representatives of both public research and higher education institutions; consequently, the approval process was delayed. In the meantime, a movement representing public research and higher education institutes, *Sauvons la recherche* (SLR), kicked off in 2003. Here, researchers chose to voice their worries to the press and to put pressure on the government regarding the design of the new Law for Research. They expressed concerns regarding some of the core measures included in this draft—like the plan to launch a National Research Agency (ANR) for the funding of research, its increasing budgetary power, and its subordinated role to the minister. Researchers were afraid that this would be detrimental to the multidisciplinary nature of public research organizations, like the CNRS, and it would allow the government, rather than the research community, to pick and choose new areas of research (Schoen et al. 2008).[13] Along the same lines, they also claimed that the increasing political focus on applied research should not overlook the relevance of basic research since these are both necessary for their mutual development. The triggering event (April/May 2004) for the beginning of a strong protest movement was the decision of the minister of research to change the status of 550 life-long research positions and transform them into fixed-term contracts. This unveiled a profound unease throughout the French research system, and it highlighted very sensitive issues such as lack of funding, excessive administrative burdens, and the inadequacy of the status of researcher as defined by the current legislation (Law 1982). On top of the reintegration of those researchers and the creation of additional university positions, the movement triggered a vivid national debate which culminated, thanks to the mediator role played by some exponents of the institutional and scientific world, with the "*États généraux de la Recherche*" (The General State for Research) in 2004. This initiative, as with previous national consultation practices, gathered researchers and citizens at both national and regional levels to make proposals for contributing to the design of a new *Pacte Pour la Recherche*. Meanwhile, internal pressures were also pushing for a reform of the research system. Indeed, in 2005, the Court of Audit released a report (*La gestion de la recherche dans les universités*) which proposed to increase the management autonomy of universities, the creation of a single evaluation authority, while introducing the idea that, to be able to compete internationally, French universities should start to gather their resources.

4.3.1 The Research Act

The Research Act bill was presented to the parliament by the minister of national education and the minister delegated for higher education and

[13] In their words: "this piloting is done by the prince, and as such the direction will change with the mood of the prince". [...] "*Ce pilotage est le fait du prince, et comme tel, le cap changera avec l'humeur du prince*" [...]
http://sauvonslarecherche.fr/spip.php?article1481

research for final approval in the spring 2006. It was committed to increase public R&D funding and the overall strategy revolved around different axes of interventions. First, it reorganized the governance of the system through the creation of an agency devoted to the rationalization of the research system evaluation (AERES *Agence d'évaluation de la recherche et de l'enseignement supérieur*) as well as an advisory council (HCST *Haut conseil pour la science et la technologie*), aimed at supporting the president of the republic on science and technology decisions. Secondly, the reform modified the landscape of the system for funding research. Traditionally public research was funded through contracts between the state and the institutions performing research activities. After the creation of the National Research Agency (*Agence nationale de la recherche*—ANR), the government was clearly moving from a mechanism of block funding toward a system of competitive funding. Moreover, it created new instruments to cluster universities on a regional basis (PRES), with the aim of reducing fragmentation (Schoen et al. 2008). The underlining logic of the PRES was that of gathering universities, *Grandes Écoles* and PROs, to enhance their visibility (size matters in the calculation of international performance rankings) and therefore increase their attractiveness in a context of high international competition.

According to Musselin and Paradeise (2009), this strategy of developing incentives for research excellence, while clustering institutions through bottom-up mechanisms eventually activated by territorial actors (e.g., presidents of universities, heads of research organizations, regions, local industries), accelerated the blurring of competences between higher education institutions and public research organizations. Overall, the instrument mix of the Research Act was characterized by a high share of low coercive instruments, designed to leave the target group enough room for maneuver in engaging with joint research activities. This was compensated by structural changes in the governance of the sector through the introduction of an evaluation agency (AERES) and a funding agency (ANR), which represents a first move toward the rationalization of the evaluation and funding system.

4.3.2 *Law on the Freedom and Responsibilities of Universities*

The following stage in the reform of the national public research system involved the transformation of university governance whose main characteristics were defined by the *Loi Farue* (1968) and the *Loi Savary* (1984).

A first attempt in the direction of enhancing the autonomy of the Higher Education Institution was proposed in 2003 by the minister for youth, national education, and research: Luc Ferry.[14] This reform was initially slowed down by the current prime minister and the president of the republic, who feared an uprising among students and academic personnel.[15] However, the issues related with the role of universities, as well as a more generalized political will to modify their internal governance structure, were still present in the national policy agenda. Indeed, a window of opportunity for promoting the reorganizations of the national higher education governance was found in conjunction with the release of the first Shanghai University rankings in 2003. Here, French universities scored lower than expected, and, as described by Musselin (2017), this event came to reinforce the initiatives that had been already ongoing, rather than triggering new ones. On the public research side, the need for more internationally competitive research systems was translated by policymakers into the Research Pact (2006), as discussed in the previous paragraph. While, for higher education institutions, the increase in public attention toward the role of universities helped to overcome the stalemate of the Ferry proposal, which culminated in the Law on the Freedom and Autonomy of Universities (2007). Another important factor supporting the approval of this reform was the election of a new president of the republic, Nicolas Sarkozy, who endorsed the reform as one of the main goals of his mandate.

Designed in combination with the 2006 Research Act, specifically regarding PRES clusters, the *Loi Relative aux Libertés et Responsabilités des Universités* (LRU Law on the Freedom and Responsibilities of Universities) reformed the governance of national higher education. First, it reduced the size of university administration councils, while increasing the role of university presidents, also in the recruitment of staff. Secondly, it introduced the possibility to recruit academic and administrative staff for temporary positions. Universities acquired greater budgetary and financial management competences, and their administration boards increasingly included external stakeholders such as representatives of

[14] This proposal must be read in light of the historical phase France was undergoing. During these years, the government managed to successfully implement the 2003 Innovation Pact while starting up the draft for the Pact for Research (2006).

[15] Which was the case, one year later, with the SLR movement in the context of the 2005 Pact for Research proposal.

local/regional authorities, entrepreneurs or executive managers of companies, and representatives of the economic and social world. However, in exchange for increased autonomy, universities had to stipulate a four-year contract (*contrat quadriennaux*) with the Ministry of Higher Education and Research for shared management of their institutional strategies.

This reform must be interpreted together with the Research Act. Indeed, the latter created a funding agency (ANR) and an evaluation agency (AERES) that increasingly undermined the role of public research organizations as funding actors (Musselin 2017). While the LRU attempted to place universities at the center of a system as main operators of public research. To some extent, policymakers aimed at counterbalancing the decreased power and resources of PROs by increasing the social and economic role of universities (especially from a locally/regionally based perspective). However, these institutions remained weaker compared to other international research institutions, and much of their research was still done in mixed research units, some of which were still controlled by the CNRS or other PROs (OECD 2014).

4.4 The Evolution of Industrial Actors: From "*Champions Nationaux*" to "*Gazelles*"

Since 1945 several actors in the French economy were consolidated into nationalized services (banks, gas, electricity). Public control of industry and finance was further deepened until the end of the 1980s when, under the Mitterand (and the later Chirach I) presidencies, facing serious fiscal constraints, the State began its privatization policy (Chandler 2014). From the beginning of the 1990s the French centralized model of state intervention (Grand Programs) started a transition toward a dismissal of the majority of its public companies (Mustar and Larédo 2002). Different cabinets tried to move away from the dominant rhetoric of supporting large companies, although some of these actors are still benefiting (albeit more indirectly) from the lion's share of public support for R&D (Eparvier et al. 2009, 21). In 1993 Chesnais wrote: *"France is a country in which there is a continuous talk about the importance of SMEs and innovation, but where, barring a few exceptions, one finds only limited evidence of their role as active components of the innovation system"* (p. 193). The policy legacy of the "Grand Programs" model was difficult to dismantle, as confirmed by both the low level of private investments in R&D and the well-pronounced state propensity to invest in specific R&D-related sectors. However, measures devoted to SMEs, along with more network-oriented approaches, were already in place in the early 1980s (see Table 3.e in the Chapter Appendix), and their activity had slowly been supplemented by other instruments. Therefore, the gradual refocusing of State intervention on industry moved toward the narrative of "new industrial policies," with initiatives focusing on community needs as major drivers for economic development.[16]

This transition was initially supported by thematic research and technological networks aimed at bringing together R&I actors at the local level while assisting enterprises in gaining familiarity with new funding instruments. The instrument mix was made up of three types of interventions:

- targeted actions supporting innovation activities (ATOUT, SOFARIS, API, Innovation Award, Support investments in key technologies, FCPI);
- targeted actions enhancing researchers' mobility and employment opportunities in SMEs (CORTECH, ARI, DRT, Post-Doc recruitment);
- development of incubation and technology transfer services (CRITT, CRT, RDT, PFT).

Its underlining logic was to boost the creation of different partnerships between local stakeholders (regional authorities, higher education institutions, PROs, public agencies), in the attempt to also involve those enterprises not targeted by traditional innovation initiatives. Here, policymakers wanted to raise the awareness of business actors while providing the necessary support needed to navigate across the intricate system of public funding. The behavioral changes—supported by instrument action—aimed at persuading local stakeholders to create collaborative networks, while relying on specific economic instruments to finance those partnerships. From a diachronic perspective, the blending of different instruments reflects patterns of selection oriented toward the adoption of very similar shapes addressed to the same target (SMEs)—a stratification pattern of tool

[16] Some examples of this new narrative are the National Research and Innovation Strategy, launched in 2009 (and replaced by National Research Strategy in 2014) and the Investments for the Future Plan adopted between 2009 and 2010 (OECD 2014).

selection (Capano and Lippi 2017)—in the attempt to trigger gradual policy changes.

However, in the case of incubators and technology transfer organizations, this selection pattern created some inconsistencies in the resulting policy mix. The regional centers for innovation and technology transfer (CRITT), created at the beginning of the 1980s, were the first local interfaces between public research and SMEs. These organizations were basically partnerships between the minister of higher education and research, on one hand, and local authorities on the other, financed by state-region contracts and focused on enhancing competences of the stakeholders involved. To monitor their activities, a few years later, policymakers created a second organization, again in the form of a partnership between institutional actors (Technological Development Networks—RDT), with the task of awarding quality labels[17] to the technology transfer services provided by the former (CRITT). With the creation of the Technological Platforms (PFT) ten years later, policymakers wanted to re-arrange the organizations involved in regional technology transfer activities. However, PFTs created an additional superstructure embracing the actors involved in local incubation activities (CRITT), to first inventory the activities in place and, second, to boost their integration. This diachronic layering of similar instruments suggests the effort of achieving coherence between instruments, without altering the distribution of power and competences among the actors already involved. This allowed policymakers to gently steer a so-called bottom-up mix toward their preferred dynamics of action. Therefore, despite an increasing attention toward SMEs and local stakeholders, the *modus operandi* of policymakers hasn't changed much, being mainly centered on an inducement of "subsidies in exchange for intended behaviors," and a still relevant role of intermediary (semi-public) organizations in the delivery of those instruments.

The increasing attention toward national innovative SMEs was accompanied by the introduction of a more supportive legal framework. The Law Dutreil (2003) introduced an instrumentation aimed at supporting business creation through the combination of administrative simplification and tax reliefs. Indeed, the instrument mix implemented during the 2000–2005 period was characterized by a combination of different expenditure instruments, blending shapes of highly coercive and centrally

[17] These were the Technological Resources Centers (CRT) as described in Table 4.5 in the Appendix of this chapter.

steered instrument shapes.[18] In this phase, policymakers moved away from the "subsidies in exchange for expected behavior" and adopted a combination of information instruments and indirect financial participation. Therefore, after a first period (1980s–1990s) of funding in exchange for expected behaviors, combined with a highly centralized instrumentation, the system supporting SME's innovation activities evolved into a policy mix providing these actors with greater autonomy, and a simplified institutional structure, while also assigning them with greater responsibilities in managing their innovation activities.

4.4.1 *The Competitive Cluster Approaches*

While negotiations for the Pact for Research were still ongoing, the minister of economy, finance, and industry approved several relevant measures for the development and support of national innovative enterprises.

First, the *pôles de compétitivité* (2004)[19]: industrial clusters promoting collaborative R&D projects between different stakeholders (OECD 2014). These networks were territorially defined by the coexistence of a highly specialized industrial base, combined with research and education potential that, through the creation of partnerships, can get access to funding under privileged conditions and rebates on corporate and social taxes (European Commission 2005). Their implementation was praised by three different expert reports. The *Ailleret* Report (2003), drafted by the National Economic and Social Council, pointed out some inconsistencies in the national R&I system, such as the lack of coherence between the interventions of local actors, the complexity of the instrument mix put into force, and the extent to which their intricate nature could hamper the effective exploitation of some services by SMEs. A second report, produced in 2004 by the agency for spatial planning and regional actions (DATAR), endorsed the development of an industrial policy based on synergies between localized agglomerations of enterprises, scientific and technological potential. Leveraging the success of these strategies in some

[18] The proximity investment fund, the Participatory Priming Law (*Prêt Participatif d'Amorçage* PPA), the Innovation Development Contract (*Contrat de Développement Innovation* CDI), the Technological fund of funds (*Fonds de Fonds Technologique* FFT), and the Gazelles programs. For more details, please refer to Table 4.6 in the Appendix of this chapter.

[19] Territorially based competitive clusters aiming at fostering the synergetic development of specific sectors, or a technology, through active partnerships between their members.

international contexts, the Blanc Report (2004)[20] also stated the need for a new national model of economic development revolving around public research organizations, universities, and entrepreneurs. The guidelines emerging from these expert reports were channeled into the strategy for the competitive clusters adopted in September 2004. Consequently, the *pôles de compétitivité* represented a clear attempt to make use of the local competences by creating an ecosystem for R&I stakeholders. Competitive clusters slowly became part of an innovative approach for state action based on a public procurement whose design was deeply inspired by the American "small business act."[21] More precisely, with the Law for the Modernization of the Economy (2009–2010), a public procurement instrument mix was introduced within the French R&I instrument portfolio. This was made up of three interventions[22] integrating (public and private) procurement provisions for SMEs, together with a system of targeted funding supporting the engagement of local industrial actors.

The SME pact created the legal framework for the development of public procurement for SMEs, basically representing a voluntary commitment made by public and private actors toward working with innovative SMEs. The *Passerelle* program, an economic tool, supported the costs for the development of innovations by SMEs while the small business act effectively established public procurement. This mix was most recently supplemented by the introduction of two different shapes of economic instruments, always targeted to help SMEs financing their innovation activities, namely: the *aide au projet d'innovation stratégique industrielle* (ISI) and the *Fond Stratégique d'Investissement* (FSI). The former supported SMEs' financing, while the latter allowed policymakers to become shareholders of the companies they were financing.

[20] This report responded to the request of the prime minister in his letter of September 30, 2003, which awaited the definition of concrete measures that could promote the establishment of competitiveness clusters.

[21] This framework was the milestone of US SMEs policies, which affirmed the need to prioritize the action of public authorities toward these actors. Other countries, such as the UK and the Netherlands, designed very similar instruments inspired by the US example (OECD 2014).

[22] The SMEs Pact, the *Passerelle Programe* and the "French-Style" Small Business act.

4.4.2 *The Tax Credit for Research*

Despite French R&I policy mix having been mainly characterized by the frequent stratification of different shapes, it is worth highlighting the case of an instrument introduced in 1983 and still operative nowadays: The Research Tax Credit (*Crédit Impôt Recherche*, CIR). This is a horizontal, non-discriminatory across sectors, fiscal instrument supporting private R&D investments by means of tax incentives calculated on their expenditures. It was introduced during the dismissal of the state's role in national industrial programs (1983), as a strategy to boost private contributions to national innovation efforts. And, at its latest stage, it was made of two credit calculation components: a volume share, determined upon the R&D expenditures incurred over one year, and an increase share calculated according to an augmentation of expenses in relation to the average expenditure over a benchmarking period.

When first introduced, the CIR was incremental in nature, the tax relief was proportional to the increase in the company's R&D expenditure compared to a benchmark period, and it was renewed by legislative provision on a multi-year basis (3–5 years).[23] Instruments recipients were dependent upon this time span for programming their innovation projects. Consequently, "the mere fact of taking into account an increasing share made this device a marginal instrument, except for companies in their first years of life and some companies constantly increasing R&D" (Larrue et al. 2006, 89). Following the release (2003) of a report drafted by the tax council criticizing the measures as being unattractive for enterprises because of its complexity, the budgetary law for 2004 modified its nature: the legislative provision of the CIR was stabilized and transformed from a multi-year to an annual basis. Moreover, it was reorganized: one component became increased based (45%), and a second was calculated on the volume of expenses (5%) and the maximum expenses ceiling was increased to 8 million euros. The latter allowed the calculation of the tax benefit according to the specificities of the project involved, rather than being dependent upon the expenses undertaken in previous years. Consequently, the credit was no longer estimated in dynamic terms but only considered the volume of research expenses. The first argument for this change was simplification: the incremental system resulted in specific calculation difficulties, and involved SMEs were often forced to hire costly specialist

[23] This consisted of 25% of the increase for a maximum ceiling of 3 million Francs.

services to "optimize" their tax declarations. Additionally, an incremental credit did little to incentivize companies whose R&D expenditure was stable over time; for example, after a previous spike (OECD 2014, 189).

With the 2006 budget law, the CIR was modified again by the introduction of the following:

- new costs as eligible expenses;[24]
- the immediate reimbursement of a research tax credit in the year of creation, and the following four, for young companies (younger than five years old);
- an increase in the benefits calculated on the volume of expenses (10%) and a reduction for those calculated on an incremental basis (40%);
- the maximum expenses ceiling was increased to 10 million euros.

Finally, also with the budget law of 2008 the tax credit was further reshaped to acquire the characteristics of the system that is still largely in place today, namely:

- a suppression of the increase-based component;
- a strengthening of the volume-based component (a tax credit of 30% of research investments will be granted for investments up to 100 million euros and of 5% beyond);
- the eligibility of all the expenses classifiable as research (the OECD Frascati Manual definition);
- the maximum expenses ceiling has been eliminated.

The evolution of this instrument reflects the changes taking place in the French R&I system at large. The first CIR reform, when the instrument was stabilized with a volume-based calculation of the expenses, took place under Francis Mer, as minister of economy, finance, and industry. He is a businessman who had a prominent role within the MEDEF, the largest employer's federation in France. In 2000 the MEDEF published a report called "Priority measures to boost innovation in France."[25] There, among

[24] (Patent protection expenses; subcontracting fees for research made by PROs, universities, or technical research centers; expenses related with employment of PhD and their contractual stabilization).

[25] "*Les mesures prioritaires pour une accélération du mouvement de l'innovation en France.*"

other claims, the organization clearly stated the need for changing the methods for calculating the credit, the adoption of the OECD definition for R&D activities (less restrictive than the French identification), together with an extension of the base for the credit's calculation.[26] Moreover, the reform took place in parallel with the implementation of cluster policies. Consequently, the re-calibration of the tax credit was also simplifying the adoption of CIR by SMEs, which were important actors in cluster development.

The second re-calibration of the instrument (2006) took place under a new minister of economy, finance, and industry (Sarkozy). This reform introduced new costs as eligible for the tax deduction, together with the possibility of subcontracting research to third-party actors (e.g., those involved in the future PRES or in the competitive clusters) as well as an increase in the ceiling of expenses. With the 2006 intervention, the tax credit included new eligible expenses related to the subcontracting of research made by PROs, universities, and other research centers. In the same period, after the Research Pact (2006), new contractual tools for grouping HEIs and PROs in geographical proximity were created (PRES). Consequently, those actors, together with the competitive clusters, were listed among the beneficiaries of the subcontracting research services supported by the tax credit. Finally, the 2008 reform, under the combination of Christine Lagarde as minister of the economy, finance, and industry and Nicolas Sarkozy (as President of the Republic), completely suppressed the increase-based calculation; it adopted OECD classification of R&D expenses (included in the Frascati Manual) and it removed any ceiling. Therefore, as also emphasized by some interviewees, it effectively seems that the evolution of the CIR has somehow taken inspiration from the suggestions made by the MEDEF report in 2000. The re-calibration of this instrument also evokes how the tool has been molded according to the needs of specific constituencies. Namely, the balance between increase and volume-based calculations (removed by the 2008 reform) awarded private actors who performed increasing innovation activities over time or new enterprises which had no previous fiscal records of innovation (which could benefit from a generous CIR during the first year). On the other hand, the shift toward a fully volume-based calculation of the CIR allowed enterprises to get a constant share of tax credit according to the amount

[26] https://www.lesechos.fr/2000/12/le-medef-souhaite-une-loi-cadre-pour-linnovation-758745 (last accessed 13 March 2022).

of innovation financed, somehow penalizing more dynamic, and less constant, innovative enterprises. However, the most relevant change was the elimination of the expense's limits. These ceilings have been growing, starting from 2003, moving up with the 2006 reform, only to be completely removed under the Sarkozy presidency reform. These changes basically paved the way for big enterprises to exploit the benefits related to the R&D credit; since before, with the limit of expenses, they could only partially benefit from it. Therefore, after the removal of the threshold, the CIR became highly convenient for big enterprises (INTERVIEWEE 15).

Data from the annual evaluation report on the CIR[27] describe how the share of SMEs (enterprises with less than 250 employees—blue area) benefiting from the credit is higher compared to the number of big enterprises (enterprises with less more than 250 employees—green area) (Fig. 4.2). On the other hand, when looking at the expenses of R&D declared, it is clear how bigger enterprises tend to represent the lion's share of the credit beneficiaries (Fig. 4.3).[28]

So, SMEs represent the largest share of private R&I performers applying for tax reduction within the CIR scheme with peaks of 86% of the total sample in 2005, 88.9% in 2012, and 89.8% in 2013. However, when looking at the amounts of tax credit gained by companies, the situation is almost flipped over, with big enterprises reaching peaks of 69.5% of the total benefits provided in 2010 and stabilizing around that amount for the rest of the period under investigation. It is also worth noticing how SMEs dropped from a value of 55% in 2005 to 21.7% in 2007, followed by a slight increase in 2008 (27.4%), which slowly increased to a final peak of 31% in 2013. This trend goes in parallel with the gradual rise of the maximum threshold for investments, which basically paved the way for a greater involvement of big enterprises.

The CIR represents an important tool of French instrumentation, and its relevance within the national R&I policy mix has increased steadily across

[27] Minister of higher education and research: http://www.enseignementsup-recherche.gouv.fr/cid49931/cir-les-statistiques.html (last access 13 August 2021).

[28] (For both Fig. 4.2 and 4.3, data are elaborated by the author according to the official statistics published by the yearly evaluation report on the Research Tax Credit (CIR) by the Minister or Higher Education and Research).

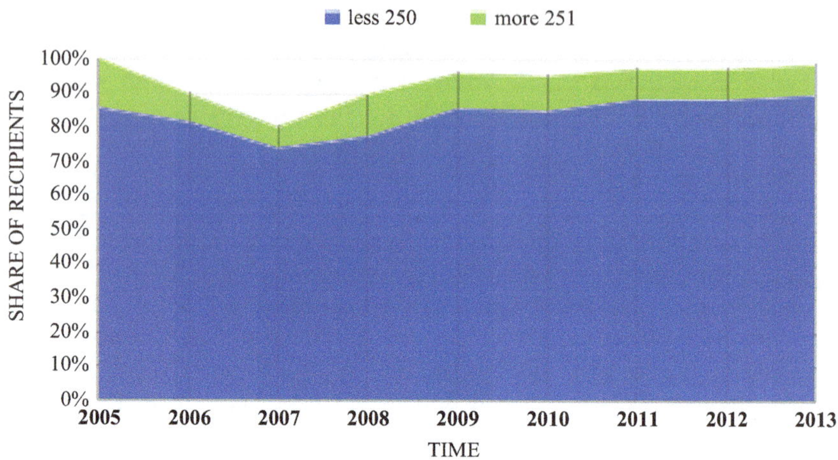

Fig. 4.2 Recipients of the Tax Credit for Research by enterprise size

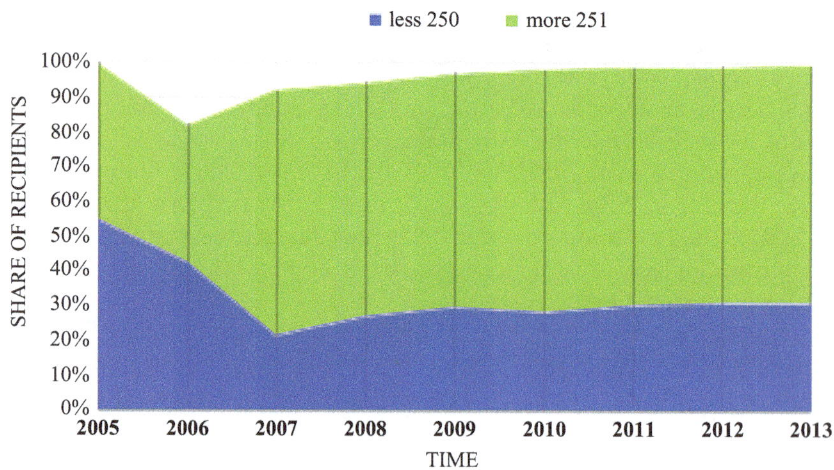

Fig. 4.3 Share of Tax Credit for Research benefits by enterprise size

years.[29] Also, foreign companies contributed to this shift to the extent that the features of the CIR were perceived as an incentive to move R&D activities to France (Ministére de l'education nationale de l'enseignment supérieur et de la Recherche 2014). This instrument did manage to survive, from 1983 until the present day, because it was created and modified by different parties in government, while never being clearly patronized by any political faction (INTERVIEWEE 13). Moreover, enterprises did like it; therefore the instrument had a small chance of being wiped away.[30] The combination between a strong constituency in support of the instrument, together with the intra party stratification process, crystalized the position of the tax credit within the broader national R&I policy mix.

4.5 Two Souls of Ministerial R&I Coordination

When external ministerial coordination practices are in place, effective R&I policymaking can be slowed down by contrasting framings on the policy issues that need to be addressed. In the French context this is characterized by a perception of the minister of higher education (MESR) as "the minister of public research organizations." So, universities and public research organizations represent the focal political interest of ministerial interventions (INTERVIEWEES 21 and 15) while the Ministry of Economy, Finance, and Industry is perceived as being primarily oriented toward helping enterprises and the innovations developed by them. The two ministries have different "constituencies," with different interests and needs. Consequently, they tend to see things differently (INTERVIEWEE 17) and therefore shape policy instruments in a different manner. This mechanism held true, and became more relevant, in the case of ministries with different sectorial competences.

[29] It represented a credit of EUR 4.5 billion in 2010, and it has represented around EUR 5 billion per year since then and could reach EUR 7 billion per year once fully operational (French Court of Auditors 2013), that is, between four and six times the amount of direct aid and around one-third of public R&D expenditure (in which it is not counted) (OECD 2014, 188).

[30] It is good to abolish fiscal niches, but within each niche there is a dog that bites (INTERVIEWEE 17).

Despite a long history of programmatic governmental efforts for coordinating national R&I policymaking competences,[31] ministers maintain strong specializations in support of their specific research domains. This pattern is consistent with the proliferation of sectorial public research organizations attached to different sectorial ministers (Mustar and Larédo 2002). The 1982 Law on Research first attempted to tackle this issue by introducing a financial division between public research organizations and their parent (functional) ministries. Major public research organizations were put under the double remit of the Ministry of Higher Education and Research and their specific functional ministers. Secondly, it introduced a budgetary platform—the Civil Budget for Research and Technological Development (BCRD)—led by the Ministry of Higher Education and Research, which merged credit allocations for public research within a unique procedure. It also covered the National Fund for Science (FNS) dedicated to fundamental research (higher education was on another budget),[32] and the Technological Research Fund (FRT) for industrial researchers and EPICs. However, the funding support for technological development of enterprises, the Fund for Enterprises Responsibilities, remained under the full responsibility of the Ministry in charge of Industry. This is symptomatic of the reduced capacity of BCRD to act as an instrument for interministerial coordination. Moreover, when a research sector was deemed as strategic for a given ministry this carried the possibility of negotiating directly with the minister of Finance. By doing so, its R&I budget appropriations entered under their full management responsibilities, thus overtaking the coordination competences of the Ministry of Higher Education and Research. This was a simple process especially for the Ministry of Industry which was integrated into the Ministry of Economy and Finance since 1997 (Cytermann 2006).

A second attempt to enhance thematic-R&I policymaking coordination focused on bridging the gap between higher education and public

[31] A first attempt was the creation of the Commissariat for Atomic Energy (CEA, 1945), together with the creation of the High Council for Scientific Research in 1954. Moreover, two years after the "*Colloques de Caen*" (a participative platform of R&I stakeholders) De Gaulle established three different organizations for coordinating and advising French R&I policymaking, the interdepartmental committee for scientific and technical research (CIRST), a deliberative decision-making body chaired by the prime minister, an advisory committee on scientific and technical research (CSRT) and a general delegation for scientific and technical research (DGRST) that was responsible for the preparation and execution of research policies.

[32] Coordinated Budget for Higher Education (BCES).

research as a part of the national reform of public management promoted by the Constitutional Bylaw on the Finance Act (LOLF) approved in 2001. This introduced a new R&I common budgeting procedure, the Interministerial Mission for Research and Higher Education (MIRES). Accordingly, R&I credits in the budget law were grouped by missions and then organized according to different programs referring to a specific policy field. Here, the minister for research was acting as a coordinator, while being the main interlocutor of both the ministries responsible for the budget, during the preparation phase of the budget law, and of the parliament, during the examination of and vote for appropriations. However, also in this case, functional ministries preserved their own budget, and competences, over the design of ministerial research strategies (OECD 2014). Consequently, the new common budgeting procedure resulted in a merely formal procedure. The programs included in the strategy mirrored sectorial research competences showing how the dialogue between ministers was actually less structured than among departments within a unique ministerial organization (Cytermann 2006). Therefore, once again, the ministry in charge of research was a king without a kingdom. It was appointed with the role of coordinator without having the necessary powers to steer sectorial ministers toward collaborative decision-making. Certainly, the consultative nature of these strategies, the weak role of some interministerial forums (e.g., CIRST), and the lack of any type of incentive for collaboration certainly hampered their success (INTERVIEWEE 17).

The National Strategy for Research and Innovation (SNRI), approved in 2008, was a further attempt to promote policymaking coordination. This consisted of a document providing long-term political orientations for national R&I policies, with the aim of identifying a common agenda among stakeholders for a period of four years (2009–2012). This was a bottom-up priority-setting process involving research, business, and civil society stakeholders, identifying three main social challenges at which national R&I policies should be directed, namely: health, environment, and ICT. Its binding powers were limited because the strategy was not set at the operational level, and it didn't allocate a budget, while being merely oriented toward the provision of thematic guidelines (OECD 2014). Indeed, the SNRI effectively took off only with the approval of the "Investment for the Future Program" (PIA) that, for its first wave of application (2010–2014), was in line with the main priorities of the SNRI (Eparvier et al. 2011).

Finally, in combination with the implementation of the PIA program, an additional committee was created, the *Commissariat Général à l'Investissement* (CGI), with the task of supervising the implementation of the plan. This organization attempted to re-centralize R&I policymaking competences under the authority of the prime minister. Indeed: *"with the CGI they didn't create coherence among the organizations involved, but rather competition between ministries and the PIA"* (INTERVIEWEE 15). The CGI represented a very powerful actor in the interpretation of the PIA, since every decision regarding the allocation of funding had to be negotiated with the committee (INTERVIEWEE 12). Indeed, it drafted contracts with the agencies in charge of its implementation (ANR; CDC; ADEME), providing specifications to supplement calls for proposals, finally ensuring the overall consistency of the investment plan through yearly reports of activities. However, despite the centralization of decision-making competences under the supervision of the prime minister, and the implementation of the plan through public agencies, organic links between ministers in the affected areas of competences were still underdeveloped.[33]

However, the interaction between different ministers has not always been characterized by inconclusive attempts at coordination. For example, when the Research and Innovation Act was approved (1999), the minister of higher education and the minister of industry were known for getting along very well (INTERVIEWEE 15). This collaboration gave birth to the most relevant French R&I policy of the 2000s, which indeed created synergies between the different "constituencies" of the two ministers, and it was supplemented and consolidated by the Innovation Plan (2003), which further consolidated these strategies. However, this is symptomatic of the fact that coordinating ministers with functional R&I competences often requires a strategy. Indeed, administrative organization can only partially explain this; it is not administrative reunification itself that makes the strategy but rather the opposite. Therefore, the selection of an external ministerial specialization pattern can rather be interpreted as symptomatic for the lack of a holistic R&I national policy strategy. Indeed, in the cases of encompassing R&I policies, as discussed in the previous paragraphs, we have witnessed the combination of either coordination between ministers or the definition of national long-term R&I strategies supported by the re-centralization of powers in the hands of a politically steered committee.

[33] For what concerns the integration of the PIA within the broader framework of ministerial programs, it is worth noticing that, after 2014, the CGI reports to the minister with delegated competencies to industries, in an effort to ensure operational cooperation with other innovation programs. Before, it was reporting to the prime minister only (OECD 2014).

4.6 Public Agencies as a Strategy for Dialogue with Strong Interests

The ecology of actors populating the French R&I public sector is highly heterogeneous. In the public research realm, there are two typologies of stakeholders: public research organizations and universities. The former, despite being characterized by different field-specialization and institutional natures, are strongly attached to their legacy of former pillars of publicly financed industrial projects. The latter represent relatively young institutions in the French context,[34] which have seen their autonomy and powers evolving across the last decades. The relationship between public research organizations (EPST—*Établissement public à caractère scientifique et technologique*/Public Institution for Science and Technology), HEIs, and the minister of higher education and research define a complex mechanism of funding streams, also including different functional ministers according to specific policy issue. Major research organizations are linked to their respective parent ministries through contracts.[35] These revolve around the agreement on general objectives which the PROs must take into consideration during the internal allocation of priorities between research teams. Similarly, universities negotiate contracts with the Ministry of Higher Education and Research, which also involve the provision of specific funding allocations. PROs have greater powers compared to universities because they bring together under a single organization different functions that in other countries tend to be arranged across different institutions: for instance, orientation, funding, execution, and evaluation of research in their respective fields (OECD 2014, 125). Therefore, the public research landscape defines an unbalanced distribution of powers and roles between different players. Indeed, PROs detain a considerable supervisory latitude in their strategic choices and internal allocation of resources, being quite influential in setting national research priorities (IGAENR 2012). Despite the institutional architecture designed for the definition of

[34] These were suppressed during the French revolution for being reorganized into faculties during the late 1960s (Musselin and Paradeise 2009). With the implementation of the Law Faure (1968) universities became public institutions of a scientific and cultural nature, and they replaced the faculties.

[35] The contractual nature of the relationship between research performers and policymakers was introduced by the 1982 Law on Research and further enhanced by the 1999 *Loi Allegre*, but their provision became compulsory only in 2007 with the approval of the Law on the Autonomy of Universities (Art.7).

the MIRES, and for the discussion of PROs' budgets, the arbitration on budgetary appropriations for research was essentially done by the operator itself (French Court of Auditors 2013). Universities appear to be weaker, fragmented entities across the national territory. Meanwhile, PROs, essentially financed by the state, represent organizations with a structured network of national laboratories, which allows them to design a more consistent national strategy (INTERVIEWEES 17–21). Such a divide is not only functional but to some extent also cultural.[36] Indeed, this cleavage was also present in the SLR movement against the 2006 Law on Research. "*[...] Initially, the SLR movement was structured around mainly PRO researchers. Then, we quickly realized that it was necessary to open the movement to universities as well, and some of them joined the protest. But I think that, already by the selection of the name ("Save the Research"), some people from the universities believed they weren't concerned,* (INTERVIEWEE 19). A push toward changes in the power balance between PROs and Universities came with the 2007 Law on the Autonomy of Universities (LRU), which introduced an evaluation culture in the management of public funding systems.[37] The underlining logic of the instrument mix was exactly to put universities at the center of the system as main operator of public research (Musselin 2017).[38] This ran in parallel to the transition from a fully block-funded research system toward a system in which funding for research was also assigned on a competitive basis. Indeed, in the same year, the National Research Agency (ANR), established in 2005,

[36] As argued by one of the experts we interviewed (INTERVIEWEE 13): *"it is necessary to understand that in France universities didn't use to be relevant institutions. Differently from other countries, the executive élites are not trained within universities. This is historically related with French Revolution, at that time universities were related with the church. With the Revolution we have created new élites detached from the church, with the creation of the Grandes Écoles; which has been enhanced during the Napoleon era, and in 1945 with the creation of ENA. 'French élites' have mainly been trained at in the Grandes Écoles, not at universities. In these institutions the goal is not to 'learn to do' but to 'learn to learn.' Therefore, French élites tend to be quite detached form research and science (except for some highly research-intensive Grandes Écoles institution e.g., École Normale Supérieure or École Polytechnique)."*

[37] This law was one of the strong suits of the Sarkozy presidency, who was minister of economy, finance, and industry during the previous legislation, at the time when the 2006 Law on Research was approved. Therefore, this might also be a reasonable interpretation for its coherence with the newly approved Law on the Autonomy of Universities.

[38] An example of this logic was the creation of PRESs (Research and Higher Education Clusters), a contractual tool for grouping higher education and PROs in geographical proximity, to pool their resources and create a unique cluster to manage research, education, and valorization activities. They represented a strategy to aggregate the different components of public research according to different themes.

became a public agency with administrative and financial decision-making autonomy under the supervision of the minister of higher education and research. It also took over research funding previously financed through the Fund for Technological Research (FRT) and the National Science Fund (FNS),[39] as well as some direct institutional funding for research laboratories (Arnold 2007; Lepori et al. 2017). Then with the implementation of the Investment for the Future Plan, the ANR became the manager of the competitive funding system financing research performers, according to the thematic priorities identified by the government.

However, researchers raised concerns regarding the increasing powers of the ANR and its direct ministerial control. There was a strong disagreement between French governmental authorities and the scientific community on who should have the power, and the authority, to identify national research priorities (Eparvier, Patrick; Turcat, Nicolas; Schoen, Antonie; Carat, Gerard; Nill 2008). At the very beginning, *"the ANR was badly perceived because there were already research operators working on their programs...but everyone was doing their own thing in a corner. Indeed, despite the research project undertaken was relevant to their fields, we noticed there were different small and divided groups according to different research fields"* (INTERVIEWEE 22). The criticisms from the research community regarding the capacity of the ANR to establish thematic priorities for French public research succeeded in triggering some changes in the governance of the organization, with the introduction of the "Alliances" (2010). These coordination groups brought together public research stakeholders, according to their research domain, to enhance cooperation in setting the programmatic priorities of the National Research Agency (Zaparucha 2010). Basically, the new role of Alliances entailed reinstating the planning function itself within the PROs, which ran counter to the previous trend of separating powers implemented by the Research and Innovation Act (OECD 2014). In that same vein, the discontent of the research community pushed the ANR to introduce a first round of "*Programmes Blanc*" ("blue-sky funding"), reserving a share of the budget allocation to non-thematic programs. Therefore, despite the introduction of a public agency for the reorganization and central direction of public research funding, the main constituencies of the sector detained a relevant bargaining power and succeeded in influencing the nature, and powers, of the newly created ANR.

[39] These were the main sources of funding subsidizing basic and applied research undertaken by public research organizations, universities, and SMEs.

Another trend symptomatic of the increasing hollowing out of ANR's power can be spotted by looking at the evolution of its budget allocations. These have been gradually increasing until 2009 (right before the introduction of the Alliances), when they leveled off and began to drop, with only a slight increase in 2016–2017, as discussed in the draft budget law for 2019.[40]

The endowment provided by the agency comes from the appropriations included in program 172 of the MIRES (see Table 6.2, and after 2010 from the PIA funding). The reduction in the allocation of funding to ANR has been initially justified by a shift toward a system of direct institutional funding allocation to PROs, as decided during the *Asisses de la recherche* 2012 by the Hollande government. On the other hand, this argument was no longer valid in recent years, to the extent that several EPSTs have experienced a reduction in their institutional funding streams.[41] Hence, the reduction of the ANR's budget allocation did most likely flow either into credits for project operators in the form of recurring grants or in a generalized reduction of funding for public research.

4.6.1 *Private R&I Performers*

After the liberalization of former state-led companies, France has increasingly supported the development of national companies toward the acquisition of a dominant position in the international market, together with a more systemic support of its network of SMEs (BCG and CM INTERNATIONAL 2008). However, as discussed in previous paragraphs of this chapter, despite the fact that targets of policy actions might have changed overtime, increasingly moving toward SMEs, the *modus operandi* of public authorities didn't seem to have changed much, being mainly centered around an inducement: "subsidies in exchange for intended behaviors."

Concerning the valorization of research activities in the industrial system and, more broadly, public support to SMEs, a public agency was in place since 1967: the ANVAR. With its 24 regional offices around the country, the agency played a double role of monitoring the evolutions and results of innovative sectors of the economy while assisting the development of SMEs. It provided support through interest-free loans for SMEs,

[40] Data from senate legislative report for the draft of the budget law 2019 http://www.senat.fr/rap/l18-147-323/l18-147-32315.html (last access 13 August 2020).

[41] As stated in the legislative report for the draft of the budget law 2017 http://www.senat.fr/rap/l16-140-325/l16-140-32514.html (last access 13 August 2020).

repayable in case of successful projects, together with consultancy services. During the 1990s, ANVAR started to focus more directly on SMEs while the government introduced (1982) a specialized financial institution also aimed at supporting SMEs: SOFARIS. This agency managed a guarantee fund coordinated by the Deposit and Consignations Bank, a public sector financial institution.[42] A third organization was also created in 1997, the Development Bank for SMEs (BDPME), which provided public guarantees for high-risk investments, therefore working in partnership with private credit system institutions.

After the creation of the competitive clusters, all these agencies were merged into a unique organization OSEO, a public holding (mainly reporting to the minister of economy and finance), with the status of a private company and a mission of public interest (Schoen et al. 2008). Its tasks were structured around three subsidiary programs, corresponding to the activities of the former organizations in place: project-based support for innovative technologies (ex-ANVAR), a loan guarantee system supporting banks and providing equity contributions to innovative SMEs (ex-SOFARIS), and the financing of business investments in partnership with other organizations (ex-BDPME) (Dosso 2014). This merger was in line with the strategic orientations defined in the 2003 Innovation Plan to transform OSEO in the main public operator for promoting innovations in SMEs (European Commission 2005). Before that, budget allocations for private R&I, especially those devoted to SMEs, were administered by three organizations in addition to the delegated regional offices related with the minister of higher education and research and the minister of economy finance and industry (the Regional Research and Technology Delegation DRRT and the Regional Division for Industry, Research, and Environment DRIRE). Therefore, through OSEO, French policymakers rationalized the structures of innovation support for SMEs, along with the implementation of competitive clusters. So, by means of this composite strategy, policymakers re-centralized the implementation of R&I interventions under the supervision of the MINEFI. In 2013, OSEO was incorporated, together with a branch of CDC and the Regional Fund for Strategic Investments, within a new public investment bank called BPI France.[43]

[42] This is a state bank funding company that is heavily involved in innovation and SMEs financing (OECD, 2014).

[43] A public institution, financing innovation activity through a different portfolio of actions, from the provision of loans and guarantees to the shadowing and consultancy of innovation projects.

The evolution of this agency represents a clear proxy for the state approach toward private performers in R&I. Indeed, through OSEO, policymakers were acting like private banks, providing out-of-market loans to innovative enterprises, with the added supplement of consultancy services. While, with BPI France, the state became a shareholder of innovative projects, being more centered around the innovation potential of enterprises. This is in line with the "public holding model" (Eparvier et al. 2011) more broadly supported by the implementation of the PIA where, through the provision of different funds, the central administration detains financial stakes in innovation organizations without the cost associated with their management. As reported by one of our interviewees (INTERVIEWEE 21), OSEO wasn't structured as a private bank, especially for what concerns prudential regulation; this provided greater freedom in managing innovative projects, at the expense of higher financial uncertainty in investments. Whereas with BPI there has been an evolution toward greater autonomy in project investment decisions at the expense of bearing greater risks in private investments in innovation activities.

4.7 Conclusions

The evolution of French R&I policy mix reflects a pattern of instrument selection oriented toward the adoption of very similar policy shapes addressed to the same target (SMEs). This clearly reflects the characteristics of what Capano and Lippi (2017) identify as the "routinization," meaning the layering of different instruments to change the direction of policy mix action and stabilize their activities into a more consistent and coherent strategy. With the 1999 Innovation Law, policymakers started to increasingly rely on their cooperation with R&I performers, as demonstrated by a majority of mixed and automatic instrument delivery structures. The strategy provided generous room for maneuver in the enactment of certain instrument shapes, while prescribing the "direction to go," because of the overall high coercive nature of the mix. With the approval of the Innovation Plan (2003) there has been a clear shift toward SMEs and start-ups as main targets of public interventions. The instrument mix was characterized by low degree of coercion and a generalized propensity toward automatic instruments. Then, the competitive cluster strategy (2005) aimed at making operational the competences developed through previous interventions. Its goal was to trigger, by mainly exploiting the set of incentives already in place, the creation of an ecosystem for stakeholders where innovations could flourish, leaving greater decisional freedom to performers. Here, policymakers attempted to debunk the long-lasting, verticalized R&I national system by fueling locally based interactions between research, education, and business. However, their combination, along with an increased relevance in public procurement for innovation, overshadows the "bottom-up" approach, providing the central State administration with greater demand-side generation power.

The underlining logic of intervention toward SMEs has often been to (financially) support private actors and to accompany them through consultancy services. A pattern of *carrots, sticks, and sermons* (Vedung 1998) supports the heterogeneous French firms, in terms of size and specialization, in their innovation activities. Ultimately, with the application of the PIA the government developed a new intervention model, in which—through its public investment bank—it contributed to equity participations by acquiring shares of innovation organizations, without being directly involved in their management (Eparvier et al. 2011). Therefore, despite some smartly designed connections among different instrument mixes, the overall evolution of national strategy seemed to be stuck in the grip between a state-led coercive approach and a more R&I performer-based strategy. Similar mechanisms are present also in the public research sector, where policymakers introduced an evaluation-based logic for funding allocation through the creation of a new research funding agency (ANR) and an evaluation agency (AERES). At the same time, universities started to cluster on a regional basis (through the PRES) in an effort to reduce fragmentation (Schoen et al. 2008), all the while investing with greater autonomy provided by the application of the 2007 university reform. Finally, with the implementation of the Investment for the Future Plan, the ANR became the manager of the competitive funding system financing research performers, according to the thematic priorities identified by the government. However, despite the initial inflow of funds and the flexibility in changing its organizational structures, the ANR has struggled to find its position in the governance of the sector. Indeed, after an initial pike in project funding for public research, the French public research sector remained sensibly weighted toward PROs, which maintain a pivotal role in the management of public funding for research (Lepori et al. 2017; OECD 2014).

To conclude, French R&I governance is characterized by important path-dependency forces, especially supported by the constituencies defending crystallized instrumentation choices. The ministerial organization of policy competences suffers from low programmatic capacity and the adoption of public agencies as intermediary actors for the management of policy instruments represents an attempt to externalize the management of conflictual policy choices. However, the combination between low policymaking coordination, public and private constituencies defending specific instrument mixes, and a routinization of similar policy instrument choices did manage to nullify many attempts at policy change.

Appendix

Encompassing Research and Innovation Strategies

(a) **The Law on Innovation and Research**

Table 4.1 Instrument mix—law on innovation and research

Instrument	*Content*	*Shapes* *Family* *Delivery* *Target*
Industrial and Commercial Business Services (SAIC)	It allowed universities to create new organizations to structure and manage their activities of research exploitation (e.g., activities for which an enterprise can request a service, valorization of research activities, general management of universities' commercial activities)	Technology transfer organization [Regulation #low coercive] [Mixed automatic] [HEIs + PROs]
Technological Research and Innovation Network (RRIT)	It clustered all the players in a technological field or industry: PROs, SMEs, universities and engineering schools, professional associations and unions, technical centers, interest groups. Bottom-up approach for deciding objectives and large degree of autonomy left to the project partners	PPPs [Information # high coercive] [Mixed automatic] [HEIs + PROs + SMEs]
National Technology Research Centers (CNRT)	Ad hoc legal public/private partnerships aimed at creating links between public/industrial research, as a vector for technology transfer. Created in the context of state regions contracts. They mainly involved large companies as industrial partners	PPPs [Information # high coercive] [Mixed automatic] [HEIs + PROs + Enterprises]

(*continued*)

Table 4.1 (continued)

Instrument	*Content*	*Shapes Family Delivery Target*
Regional Incubators	Consulting structures in support of technologically based firms through different services: training, advice, and funding system	Consulting service (Incubators) [Information # high coercive] [Low automatic] [HEIs + PROs]
Co-investments in Fund for Young Enterprises	It took up minority participations in young technological enterprises, or together with the investment funds established in the European Union. It intervened under the same conditions as private investors. CDC and the France state were the major underwriters	Equity participation [Low automatic] [Expenditure # high coercive] [Start-ups]
Seed Capital Fund for Development	Seed-capital funds on major areas of technology at the national level, with the partnership of public research institutions and private investors. Through regional incubator structures it allowed the development of regional seed-capital funds to invest 75% of it in enterprises linked with public research. CDC was the major investor with 30% of total funding	Cash advance [Expenditure # coercive] [Low automatic] [Start-ups]
Tax Credit for Research Expenses	Extension for expenses associated with researchers working directly in R&D projects	Tax reduction [Fiscal # medium coercion] [High automatic] [SMEs]
National Competition for the Creation of Technology-based Firms	National competition open to anyone willing to set up a new technology-based firm. It supported the best projects through a grant and by providing services to newly created firms	Competitive funding [Expenditure # low coercion] [Low automatic] [SMEs+ researchers]

(b) The Innovation Plan

Table 4.2 Instrument mix—innovation plan

Instruments	*Content*	*Shapes* *Family* *Delivery* *Target*
CIFRE Convention	It supported the recruitment of students and researchers by private enterprises to undergo joint applied research projects supervised by enterprises and public universities/laboratories	Targeted funding [Low automatic] [Expenditure # low coercion] [SMEs]
Supports to Projects by Young Innovative Companies	It helped young and highly innovative firms to overcome the first years of existence by providing tax credit in favor of R&D investments	Tax exemption [High automatic] [Fiscal # low coercion] [Start-ups]
Unipersonal Risk Investment Company (SUIP)	It aimed at mobilizing individual investors (*business angels*) by granting corporate tax exemptions and income tax exemptions (for a fixed period)	Tax exemption [High automatic] [Fiscal # low coercion] [Start-ups]
Entrepreneurial House	It allowed higher education institutions to develop offices to open up universities to the business world (promotion of entrepreneurial culture among students)	Targeted funding [Expenditure # low coercion] [Mixed automatic] [HEIs+ PROs]
Industrial Property Pre-diagnosis	It aimed at assessing enterprise needs in terms of industrial property rights through consultations with experts	Consulting service [Information # high coercive] [Low automatic] [SMEs]
Tax Credit for Research Expenses	Extension of the exonerations granted with the 1999 Innovation Act	Tax reduction [Fiscal # medium coercion] [High automatic] [SMEs]

(c) The Investment for the Future Plan

Table 4.3 Instrument mix—investment for the future plan

Instruments	*Contents*	*Shapes* *Family* *Delivery* *Target*
Excellence Initiatives IDEX	It aimed at connecting, on a territorial basis, higher education institutions, *Grandes Écoles,* and research institutions recognized for their scientific and pedagogical excellence. They aimed at ensuring their visibility and attractiveness on an international scale (also for the economic actors)	Competitive funding [Expenditure # low coercion] [Low automatic] [HEIs+PROs]
Health and Biotechnology Programs	It aimed at reinforcing the expertise of French research in the health and biotechnology sector, by funding research in different subfields (e.g., bioinformatics, Nano-biotechnologies, and various "cohorts")	Competitive funding [Expenditure # low coercion] [Low automatic] [HEIs+PROs]
Technological Research Institutes (IRT)	It supported the creation of four to six technological research institutes, within existing French campuses. These institutes, through public-private partnerships on research, innovation, and education reinforced existing competitiveness clusters, in order to help the country reach international level in diverse economic fields and thus create growth and jobs	Targeted funding [Expenditure # low coercion] [Low automatic] [HEIs+PROs+SMEs]
Hospital-University Institutes IHU	It aims at supporting the creation of foundations associating one university, one hospital, and several research institutes on an excellence program in the field of care, crosscutting research, and clinical and applied research	Targeted funding [Expenditure # low coercion] [Low automatic] [HEIs+PROs]
Technological Transfer Acceleration Companies SAAT	It supported technology transfer acceleration companies. These will be mostly owned by a consortium of HEI and research centers. Their mission is to focus on the valorization of public research through innovative strategy with respect to patents, industrial partnerships, creation of SMEs, and researchers' mobility. They will act as service providers in the field of research valorization for their shareholders and other potential clients	PPPs [Information # high coercion] [Mixed automatic] [HEIs+PROs]

(*continued*)

Table 4.3 (continued)

Instruments	*Contents*	*Shapes* *Family* *Delivery* *Target*
Excellence Laboratories LABEX	It aimed at selecting excellent research laboratories and providing them with the financial means to compete with international research institutions, attract internationally recognized researchers, and perform high-level research and education programs	Competitive funding [Expenditure # low coercion] [Low automatic] [HEIs+PROs]
Excellence Institutes in the Field of Carbon-Free Energies IEED	It aimed to create five to ten institutes (IEED) on campuses with international visibility. Each of them could benefit from public financial support, targeted investments, and support in long-term strategy development. The overall objective is to build up several campuses with the capacity to aggregate public and private investments around promising research projects on energy efficiency	Targeted funding [Expenditure # low coercion] [Low automatic] [HEIs+PROs]
Excellence Equipment EQUIPEX	It aimed at investing in infrastructures in view of boosting the French state research and accelerating research breakthroughs, at reducing the delay in the acquisition of medium-sized equipment, especially those that fit with the national strategy for research an innovation	Targeted funding [Expenditure # low coercion] [Low automatic] [HEIs+PROs]
France Brevet	It was a public enterprise (owned by the state and CDC) serving the valorization and protection of French technological innovation. Its mission is to support companies in valuing their innovations by structuring their intellectual property and by defending it around the world	Equity participation [Expenditure # high coercion] [Low automatic] [HEIs+SMEs]

A cohort consists in following for several years, or decades, a population of subjects, healthy or sick, in order to accumulate reliable knowledge on their health. It can help improve the prevention, diagnosis, and treatment of a wide range of serious and life-threatening diseases (e.g., cancers, heart diseases, diabetes, and arthritis)

See http://www.enseignementsup-recherche.gouv.fr/cid51358/cohortes.html

The Research Act

Table 4.4 Instrument mix—research act

Instruments	*Content*	*Shapes* *Family* *Delivery* *Target*
Research and Higher Education Clusters PRES	Contractual tool for grouping higher education/ PROs in geographical proximity (district, department, region) to pool their resources and create a unique cluster to manage research, education, and valorization activities	Contract [Regulation # low coercion] [Mixed automatic] [HEIs+PROs]
Thematic Advanced Research Networks RTRA	Funding structure for research units in geographical proximity, able to gather high-level researchers working on a specific thematic field and interact with local industrial clusters	Targeted funding [Expenditure # low coercion] [Low automatic] [HEIs+PROs]
Carnot Award	It awarded (with funding) a limited number of public research entities or private research organizations with general interest goals, for their implication with the socio-economic partners. (These are supposed to be the French counterpart of the German Fraunhofer)	Competitive funding [Expenditure # low coercion] [Low automatic] [HEIs+PROs]
Mobilizing Programs for Industrial Innovation PMII	It aimed at creating and supporting consortia between enterprises, research laboratories, and SMEs devoted to the creation of highly innovative and commercially viable products (within 5–10 years)	PPPs [Information # high coercion] [Mixed automatic] [HEIs+PROs]
Agency for the Evaluation of Research and Higher Education AERES	This agency encompassed all the previous agencies in charge of research evaluation. It evaluated research institutions, research teams, and education institutions, allowing to concentrate assessment and evaluation of public research and teaching institutions within one agency	Agency for evaluation [Regulation # high coercive] [Low automatic] [HEIs+PROs]

(*continued*)

Table 4.4 (continued)

Instruments	*Content*	*Shapes Family Delivery Target*
High Council for Science and Technology HCST	It advised the president of the republic and the government in matters related with science and technology. The members were highly distinguished personalities appointed directly by the president of the republic for four years	Advisory committee [Regulation # low coercive] [Low automatic] [HEIs+PROs]
National Agency for Research ANR	It administered competitive research funding allocation. It took over the support action previously conducted with the technology research fund (FRT *fond de la recherche technologique*) and national science fund (FNS *fond national de la recherche*)	Funding agency [Regulation # high coercive] [Low automatic] [HEIs+PROs]

From the "Champions Nationaux" to the "Gazelles"

Table 4.5 Business-oriented policy mix I (1980s–1990s)

Instruments	*Year*	*Content*	*Shapes Family Delivery Target*
Support to Innovative Projects (API)	1979	It provided zero-interest-rate loans to SMEs to develop innovative projects	Soft loan [Expenditure # high coercive] [Low automatic] [SMEs]
SOFARIS (*Société Française de garantie des financements des PME*)	1982	The scheme aimed at providing economic guarantees to entrepreneurs who did not have easy access to bank loans in order to develop their SME's business	Economic guarantee [Expenditure # medium coercive] [Low automatic] [SMEs]

(*continued*)

Table 4.5 (continued)

Instruments	*Year*	*Content*	*Shapes Family Delivery Target*
Technology Diffusion within SMEs (ATOUT)	1984	Targeted funding to support SMEs' projects aimed at improving their technological level (technical and commercial feasibility studies, pre-competitive research in priority sectors)	Targeted funding [Expenditure # low coercion] [Low automatic] [SMEs]
Innovation Award	1991	It aimed at increasing general SMEs' awareness of innovation. Specifically, it rewarded SMEs and research institutes which had successfully used patents for business or innovation development	Competitive funding [Expenditure # low coercion] [Low automatic] [HEIs+ PROs + SMEs]
Key Technologies Development	1996	Targeted funding aimed at supporting the development of key technological areas in SMEs	Targeted funding [Expenditure # low coercion] [Low automatic] [HEIs+ PROs + SMEs]
Mutual Fund for Innovation (FCPI)	1997	Risk mutualization instrument to promote investments in innovative SMEs by buying funding shares in exchange for fiscal benefits	Equity participation [Expenditure # high coercion] [Mixed automatic] [SMEs]
Support for the Recruitment of Technicians on Innovative Projects (CORTECHS)	1988	Targeted funding aiming to associate researchers with enterprises through the development of a one-year research project to stimulate innovation within SMEs	Targeted funding [Expenditure # low coercion] [Low automatic] [SMEs]

(*continued*)

Table 4.5 (continued)

Instruments	*Year*	*Content*	*Shapes Family Delivery Target*
Support to Recruitment for Innovation (ARI)	1988	Targeted funding aiming at supporting SMEs to reinforce their R&D personnel and resources by hiring a post-graduate/undergraduate	Targeted funding [Expenditure # low coercion] [Low automatic] [SMEs]
Technological Research Diploma (DRT)	1997	Targeted funding aiming at supporting SMEs to reinforce their R&D personnel, by allowing young engineers to take advanced degree diploma while working in an SME	Targeted funding [Expenditure # low coercion] [Low automatic] [SMEs]
Post-doc Recruitment	1998	Targeted funding aiming at supporting SMEs to reinforce their R&D personnel by hiring post-doc researchers	Targeted funding [Expenditure # low coercion] [Low automatic] [SMEs]
Regional Center for Innovation and Technology Transfer (CRITT)	1980	Partnership between the minister of higher education and research and local authorities, providing SMEs with technological services Local interfaces between public research business, adopting HEIs and PROs competences to increase technological level of SMEs	Consulting service (incubators) [Information # high coercive] [Mixed automatic] [HEIs + PROs + SMEs]
Technological Resources Centers (CRT)	1996	Label (jointly assigned by the minister of higher education and research and the minister of economy, finance, and industry) intended to certify the good quality of the services provided by different CRITT	Label (quality standard) [Regulation # medium] [Low automatic] [HEIs + PROs + SMEs]

(*continued*)

Table 4.5 (continued)

Instruments	*Year*	*Content*	*Shapes Family Delivery Target*
Technological Development Network (RDT)	1990	Incubators associating an informal network of different institutional actors working in the field of technology transfer and SMEs support at the local level (DRRT, DRIRE, ANVAR, CRITT). They have the ultimate task to monitor CRITT	PPPs [Information# coercive] [Mixed automatic] [SMEs]
Technology Platforms (PFT)	2000	They form a network of higher education institutions, engineering schools, universities, and CRITT, providing access to equipment, training, and expertise to SMEs	PPPs [Information# coercive] [Mixed automatic] [HEIs + PROs + SMEs]

The DRRT (***Délégation régionale à la recherche et à la technologie***) and the DRIRE (***Direction régionale de l'industrie, de la recherche et de l'environnement***) are regional delegated offices, respectively, of the Ministry of Research and the Ministry of Economy, Finance, and Industry. These decentralized departments were created in 1983, and they were charged with the regional implementation R&I national policies

Table 4.6 Business-oriented policy mix II (2000s)

Instruments	*Year*	*Contents*	*Shapes Family Delivery Target*
Proximity Investment Fund	2003	Investments fund for investments in SMEs in a specific regional zone	Investment fund [Regulation # medium coercion] [Low automatic] [SMEs]
Participatory Priming Loan (PPA)	2005	OSEO loans funding early stage of innovation projects in SMEs	Loans [Expenditure # high coercion] [Low automatic] [SMEs]
Innovation Development Contract (CDI)	2005	OSEO loans funding mature (>3 years old) SMEs innovation projects in SMEs	Loans [Expenditure # high coercion] [Low automatic] [SMEs]

(*continued*)

Table 4.6 (continued)

Instruments	*Year*	*Contents*	*Shapes* *Family* *Delivery* *Target*
Technological Fund of Funds (FFT)	2005	Public equity participation in venture funds investing in technological enterprises	Equity participation [Expenditure # high coercion] [Low automatic] [SMEs]
Gazelles	2006	Support to promote venture and development capital for gazelles (most fast-growing SMEs)	Tax exemption [Fiscal # low coercion] [High automatic] [SMEs]

(d) **The Cluster Approaches**

Table 4.7 Cluster I—instrument mix

Instruments		*Contents*	*Shapes* *Family* *Delivery* *Target*
SME Pact	2005	It established that public agencies, large private companies, *pôles de compétitivité* can create procurement provisions targeted to SMEs	Public procurement [Expenditure # high coercive] [High automatic] [SMEs]
Passerelle Program	2007	Targeted funding for SMEs aimed at supporting tripartite financing for the development of the SME pact procurement	Targeted funding [Expenditure # low coercive] [Low automatic] [SMEs]
"French-Style" Small Business Act	2008	Allows SMEs preferential access to public procurement (assign them 15% of the average yearly share of their high technology market, R&D, and technological studies)	Public procurement [Expenditure # high coercive] [High automatic] [SMEs]

Table 4.8 Cluster II—instrument mix

Instruments		*Contents*	*Shapes* *Family* *Delivery* *Target*
Support to Innovative Industrial Strategic Project ISI	2008	Targeted funding devoted to help SMEs with high growth potential to develop disruptive innovation as a part of a collaborative project involving other companies/competence centers	Targeted funding [Expenditure # low coercive] [Low automatic] [SMEs]
Strategic Investment Fund FSI	2009	Public equity participation as minority shares in French companies carrying out industrial projects creating economic benefit and competitiveness	Equity participation [Expenditure # medium] [Low automatic] [SMEs]

Bibliography

Arnold, Erik. 2007. *CREST 3% OMC Third Cycle Policy Mix Peer Review Country Report France.*

BCG & CM INTERNATIONAL. 2008. *Evaluation Des Pôles de Compétitivité, Synthèse Du Rapport d'évaluation.*

Berger, Suzanne. 2016. Rapport à le Secrétaire d'Etat à l'Enseignment Supérieur et à la Recherche; le Ministre de l'Economie, de l'industrie et du Numérique *Reforms in the French Industrial Ecosistem.*

Capano, Giliberto, and Andrea Lippi. 2017. How Policy Instruments Are Chosen: Patterns of Decision Makers' Choices. *Policy Sciences* 50 (2): 269–293.

Chandler, J. A. 2014. France. In *Comparative Public Administration*. London: Routledge.

Cytermann, Jean-Richard. 2006. L'architecture de la loi organique relative aux lois de finances (LOLF) dans les domaines de l'éducation et de la recherche: choix politiques ou choix techniques? *Revue française d'administration publique* 117 (1): 85–93.

Dosso, Mafini. 2014. Restructuring in France's Innovation System: From the Mission-Oriented Model to a Systemic Approach of Innovation. *LEM Papers Series.*

Eparvier, Patrick; Turcat, Nicolas; Schoen, Antonie; Carat, Gerard; Nill, Jan. 2008. *ERAWATCH Country Report 2008. An Assessment of Research System and Policies France.*

Eparvier, Patrick, Flora Giarracca, and Léonor Rivoire. 2009. *ERAWATCH Country Report 2009: France.*

Eparvier, Patrick, Olivier Mallet, and Léonor Rivoire. 2011. *ERAWATCH Country Reports 2011: France.*

European Commission. 2000. *European Trend Chart on Innovation - Country Report France.*

———. 2002. *EUROPEAN TREND CHART Theme-Specific Country Report: France.*

———. 2004. *European Trend Chart on Innovation Annual Innovation Policy Trends and Appraisal Report: France.*

———. 2005. *European Trend Chart on Innovation - Annual Innovation Policy Trends and Appraisal Report: France.*

French Court of Auditors. 2013. *Le Financement Public de La Recherche, Un Enjeu National.*

IGAENR. 2012. *Etude Des Mécanismes d'allocation Des Moyens Humains et Financiers Aux Unités de Recherche Par Les Organismes de Recherche.* Direction de l'information légale et administrative.

Larrue, Philippe, Patrick Eparvier, and Sophie Bussillet. 2006. Etude de l'impact Du CrÉdit Impôt Recherche.

Lepori, Benedetto, Emanuela Reale, and Emilia Primieri. 2017. *Public Funding Country Profile France Annex 11.*

Ministére de l'education nationale de l'enseignment supérieur et de la Recherche. 2014. *Développement et Impact Du Crédit d'impôt Recherche : 1983–2011.*

Musselin, Christine (1958-....). 2017. *La Grande Course Des UniversitÉs.* ed. Sciences Po Presse.

Musselin, Christine, and Catherine Paradeise. 2009. France: From Incremental Tansitions to Institutional Change. In *University Governance: Western European Comparative Perspectives*, ed. Catherine Paradeise, Emanuela Reale, Ivar Bleiklie, and Ewan Ferlie, 21–49. Springer Verlag.

Mustar, Philippe, and Philippe Larédo. 2002. Innovation and Research Policy in France (1980–2000) or the Disappearance of the Colbertist State. *Research Policy* 31 (1): 55–72.

OECD. 2014. *Reviews of Innovation Policy: France 2014.* OECD.

Schoen, Antonie, Gerard Carat, and Nill Jan. 2008. *ERAWATCH Analytical Country Report 2007: France.*

Vedung, Evert. 1998. Policy Instruments: Typologies and Theories. In *Carrots, Sticks and Sermons: Policy Instruments and Their Evaluation*, ed. E. Vedung, M.L. Bemelmans-Videc, and R.C. Rist, 21–58. New Brunswick: Transaction Publishers.

Zaparucha, Elisabeth. 2010. *ERAWATCH Country Report 2010: France.*

CHAPTER 5

Research and Innovation Policy Design in Italy

The Italian innovation system has historically relied on the coordination between public research institutions and firms, which use to be predominantly state-owned entities. Similar to that of France, this economic structure was combined with a prominent role played by the public sector, both as a research producer and as a performer of innovation activities.

Like other Western European countries, Italy experienced a wave of privatization of many formerly public-owned enterprises guided, among other reasons, by the need to reduce national public debt. At the end of the twentieth century, in parallel with this privatization process, the country was also marked by the beginning of a process of public service modernization. This transition, led by means of the so-called *Bassanini reform*, involved the whole national administrative system by redefining the mission of several public institutions (Capano and Gualmini 2011). The implementation of this reform cut across different cabinet periods and involved different sectors for public administration.

In the following decade, the system was focused on the implementation of major structural reforms in the public research and higher education sectors. Meanwhile, on the industrial innovation side, policymakers shifted their attention toward the innovative potential of small and medium enterprises (SMEs) and their network-based local structures in the territory. National public finances were put under stress during the second semester of 2011 when Italy was caught in a financial storm, as a backlash of the

C. Acciai, *Policy Design for Research and Innovation*, International Series on Public Policy,
https://doi.org/10.1007/978-3-031-36628-4_5

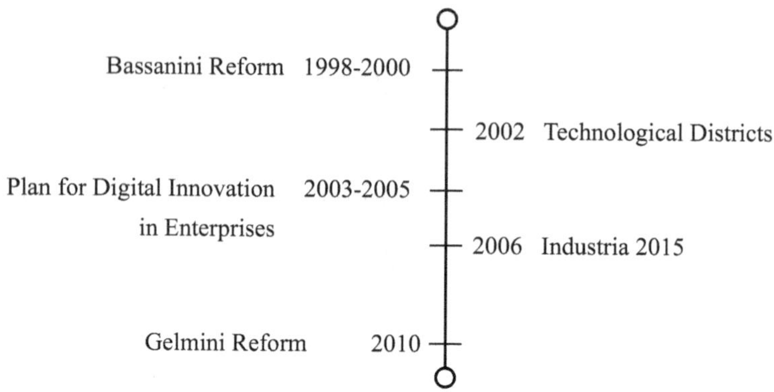

Fig. 5.1 The chronology of R&I policies (Italy)

2008 global financial crisis, in parallel with the worsening of the sovereign debt crisis in Europe. As a reaction, starting in 2012, the government concentrated its policies on public expenses reforms and on actions for promoting growth, including measures regarding R&D funding and the streamlining of main research funds. This was also in line with a generalized reduction of funding in this sector (Nascia and Pianta 2013). The most important policy events shaping national R&I policies and those analyzed in this chapter are described in Fig. 5.1.

The Italian innovation system is characterized by an industrial specialization model based on low-to-medium technological sectors, relying on low internationalized SMEs. The system suffers from a low propensity of enterprises to invest in R&I (Gallo and Silva 2006) and a more generalized lack of coordination between the field of scientific research and industry (European Commission 2001). Despite the systematic attempts to link the public and private research systems, the two sectors are still fairly separated. The former is mainly focused on research areas detached from the needs of industries; while these actors are characterized by a generalized lack of demand for scientific and technological research (European Commission 2004). The national R&I instrumentation looks highly polarized, with business-friendly innovation initiatives on the one side and a "residual" instrumentation for higher education institutions and public research organizations on the other (Nascia et al. 2012).

As discussed in previous chapters, this research adopts a broad perspective on the study of R&I policies. Therefore, as with the French case study, this chapter differentiates between policy strategies that attempted to adopt an encompassing, holistic-oriented approach (Borrás and Edquist 2019), versus those which were focused on either the research or the innovation component. To fully comprehend the features of the different R&I instrument mixes implemented across different strategies, we invite the reader to look at the supplementary material in the appendix of this chapter. Here, we decompose each specific R&I intervention according to its instrument mix components, which are then described and classified according to the proposed R&I instrument typology and the target to whom they are addressed.

The chapter is organized as follows. First, it will introduce the features of the national R&I governance structure; then, it will focus on the instrumentation and the political process characterizing the policy design of encompassing R&I policy strategies. Then, it will move on with a discussion on the reforms characterizing the public research sector, and the more general evolution of the national higher education institutions. Section 5.4 discusses the evolution of national instrumentation addressed to the industrial sectors. The final two sections of this chapter describe the national governance of R&I policy with a specific focus on the public and private actors involved in the sector.

5.1 The National Governance Structure

The Italian R&I sector is characterized by a highly centralized governance structure revolving around two core institutions: the Ministry of Higher Education and Research (MIUR) and the Ministry for Economic Development (MISE). Hence, the governance system enjoys a stable center-of-government structure, via the development of multiannual national research plans which are, however, matched with scarce monitoring procedures and uncertainties related to the availability of resources (Potì and Reale 2011). Important contributions to the evolution of the sector can also be found in sectorial ministries (e.g., Health, Agriculture, and Environment), who respectively detain public research organizations under their control (the National Institute of Health, *Istututo Superiore di Sanità*, and the Council for Agricultural Research and Analysis of Agricultural Economy). The interministerial committee of economic planning (CIPE) has coordination functions in the field of planning and national economic policy. Its main tasks include evaluating national strategic guidelines and economic strategies, including policy measures fostering research in science and technology as well as the annual allocation of resources devoted to R&I (European Commission 2006).

Public research activities are primarily financed by the minister of higher education and are concentrated in higher education institutions and public research organizations. This duality in the realm of public research requires some contextualization and discussion on the historical roots of this divide. The National Council for Research (CNR), one of the main PROs in the country, was established in 1923, and with the creation of the minister for higher education and research, more than 60 years after its foundation, the institution lost part of its funding autonomy and had to be moved under the remit of the newly created ministry. Historically, the council had overseen the coordination and the funding support of national research policies. In 1989, the coordination task was moved to the newly funded minister of higher education and research, while part of its funding activities was terminated some years later in the framework of a reform on public research (which happened in 2003). However, the National Council for Research remains the major national public research organization (Reale and Morettini 2017) supporting scientific and technological research across different fields. It operates through an autonomous development plan defining guidelines and priorities of the organization's strategy in coherence with the national strategies (Coccia and Rolfo 2010).

The Ministry of Higher Education and Research is the main funding partner of the research council, in addition to project-based missions with different functional ministers. Together with the council, other public research organizations are supervised by the MIUR, such as the Italian Space Agency (ASI) and the National Institute for Nuclear Physics (INFN), the most relevant in terms of shared funding. Moreover, there are also public research bodies under the specific control of functional ministers; among the biggest are:

- the National Agency for New Technologies, Energy, and Sustainable Economic Development (ENEA), under the control of the minister of economic development and involved in research concerning energy, environment, and new technologies;
- the National Institute of Health (ISS) under the supervision of the minister of health and involved in research concerning public health;
- the Council for Agricultural Research and Analysis of Agricultural Economy, under supervision of the minister of environment and involved in research concerning agri-food related themes.

The governance of the sector doesn't provide any institutionalized platform for dialogue among them. Therefore, their activity relies on the general guidelines defined at the national level (see below for more details on the national research program) and by their own functional ministers' policy strategies.

The national higher education system is made up of 97 higher education institutions, of which 67 are State universities.[1] Until 1994 the system of funding for State universities was dependent upon governmental decisions, and the allocation of research funds came together with restrictions on how these funds should be spent and distributed within the internal governance of the universities. The system was reformed by the Finance Act of 1994, which introduced the Ordinary Financing Fund (FFO): a lump-sum fund yearly assigned by the national budgetary law and devoted to cover universities' operating costs. This provided higher education institutions with greater freedom to decide how to spend these funds. Despite the similar research-based nature of public research organizations and higher education institutions, there is no institutionalized system of cooperation between them—in contrast to what happens in France with the mixed research units ("*unite mixte de recherche*").

The Italian industrial system is largely made up of SMEs, which tend to be mainly family-run businesses (Pierantozzi 2007). Their development, and proliferation, revolved around a specific organization of the productive system, characterized by geographical agglomerations of enterprises specialized in a related sector of activity: the industrial district. These are homogeneous production systems with a concentration of industrial companies, of small and medium size, with high productive specialization. Since the beginning of the 1980s, the organization of productive systems in traditional sectors (so-called made in Italy) in Northern regions had progressively evolved toward an organization based on industrial districts. Over the years, these expanded and consolidated, also in other regions, thanks to the successful mix provided by a combination of flexibility, usually related to the family ownership industrial structure, and the interactions with other firms in the district (M. Di Maio 2014). The national innovation system consists of three groups of organizations: universities and PROs, governmental institutions, and enterprises. All of them display a limited degree of interaction with each other, which indeed is considered

[1] https://www.miur.gov.it/web/guest/istituzioni-universitarie-accreditate (last access, October 2022).

as one of the main weaknesses of the national system of innovation (Malerba 1993).

Italy pursued an external ministerial specialization strategy; hence, R&I policymaking competences are decoupled among functionally homogeneous ministerial organizations. So, the governance of the sector hinges upon two main actors: the minister of higher education and research (MIUR) and the minister for economic development (MISE), who retain shared competences in the policy design of R&I instrument mixes.

5.2 Encompassing R&I Strategies

The political and social relevance of public research and innovation issues in the Italian case can be traced back to the end of World War II. The first reforms focused on the reorganization of public research institutions, with the creation of different institutions from departments formerly included in the National Council for Research, like the National Institute of Nuclear Physics (1945). The governance of the sector used to be characterized by a minister, without a related institutional budget, with tasks of supervision and control over the sector. In this system, the national council had a leadership role, as both research performer and manager of national R&I instruments; namely, mission-oriented research funding (so-called finalized project) and strategic projects (Potì and Reale 2006; Sirilli 2010).[2]

Technological innovation in the industrial sector was supported with the Special Applied Research Fund (FSRA) (l.1089/68),[3] created at the end of the 1960s. This consisted of a revolving fund financing research focused on economic activities. The targets were industrial enterprises, research companies, and public economic entities. It aimed at incentivizing research projects autonomously presented by R&I performers, together with a special attention toward creating networks between industrial actors, by making it easier to create a consortium of firms.

However, the first clear reference to innovation issues in the country can only be traced back to 1982, with the creation of the Fund for

[2] Finalized supported knowledge developments in specific research sectors, selected by the National Research Council, aimed at universities, public research organizations, and firms. Meanwhile, strategic projects, which ended in 2002, were collaborative research projects at the national level on priority research themes identified by the national council. These projects were evaluated by external experts selected by the council, also the main recipient of these funds.

[3] Substituted by the Fund for Applied Research (FAR) in 1999 (d.lgs. 297/99).

Technological Innovation (FIT) (l.46/82).[4] This instrument supported research activities for technological innovation and pre-competitive developments. This was combined with the allocation of further resources to the Special Applied Research Fund, together with the inclusion of a targeted funding instrument, allowing the minister for industry to devote an amount of the fund to specific (mission-oriented) research projects.

On the public research side, the first Ministry of Higher Education and Research was established in 1989. This triggered a reorganization of the sector, with a re-centralization of the national governance in the sector at the mercy of the newly created institution. This process culminated in a reorganization of national R&I governance in the broader framework of the national reform of public administration (*Bassanini* reform package). The first step of the reform (the so-called *Bassanini I Law*[5]) focused on evaluating the existing science and technology structures. The analysis highlighted how the governance of the Italian R&I sector was affected by a weak systematic approach, the lack of assessment, and foresight activities and by weak planning capacity in elaborating strategic programs (European Commission 1999). The analysis of the minister of higher education and research[6] identified five steps for the reform of the sector:

- delegation of some competences to regions (e.g., incentives in favor of innovation and the realization of interventions related with European funds d.lgs. 112/98);
- design of a new governance system for public research (d.lgs. 204/1998);
- reorganization of public research institutions (d.lgs. 19/99; d. lgs.27/99; d.lgs. 36/99);
- revision of the incentives supporting industrial research and technological innovation (d.lgs. 297/99; DM MIUR 593/00; DM MISE 16/01/01);

[4] The Law on Incentives for Technological Innovation was solicited by Italian industrial groups, such as FIAT which was at the forefront in that period (Gallo 2016, 18).

[5] (Law 59/1997), which was part of a broader set of norms (l. 127/97; l.1919/98), so-called *Bassanini* reforms, aimed at transforming and simplifying the public administration system, while decentralizing some tasks to regional and local authorities.

[6] Report to the Chambers of the Ministry of University and Research ("*Relazione alle camera del Ministro dell'Università e della Ricerca*").

- creation of a public agency "*Sviluppo Italia*" (d.lgs. 1/1999) with the scope of promoting productive activities, attracting investments, and developing demands for innovation.

These guidelines have been slowly implemented through a highly fragmented package of reforms, as summarized in Table 5.1 (Appendix).

The first instrument mix restructured the division of competence between regions and the central state for what concerns R&I funding. This entailed combining existing R&I incentives for local actors into a unique fund, the single regional fund, while assigning competences for their management to regional actors. In the framework of this reorganization was also created a new instrument addressed to SMEs: the "one stop shops for enterprise." This tool was basically a consulting service provided by municipalities to SMEs focusing on simplifying administrative burdens, while also providing consultancy on industrial activities and corporate support (European Commission 2001). The second step of the reform process redefined a new governance for the national system of R&I. The instruments came together with a reorganization of decision-making structures, with the inclusion of R&I committees in the related ministries[7] and the introduction of an evaluation system for the public research sector (Sirilli 2010). In this context of reforms, policymakers attempted to synchronize national R&I policy design to the design on national economic strategies by introducing the National Research Program. Indeed, the procedure to approve the plan was combined with national economic planning and forecasting procedures under the remit of the Interministerial Committee for Economic Planning (CIPE).

A further reorganization push was driven by the revision of the incentives supporting industrial research and technological innovation (d.lgs. 297/99). This reform combined the existent R&I instruments inside a unique legislation: the package of incentives devoted to industrial research. Further, it defined the competence division between the Ministry of Economic Development and the Ministry of Higher Education and Research, with their related funding availability.

[7] The Committee of Experts for Research Policies (CEPR), the Steering Committee for the Evaluation of Research (CIVR) within the MIUR, the Research and Training Commission within the CIPE, and the creation of technical observatories for R&I issues within both the MIUR and the MISE (European Commission 2002).

The Fund for Technological Innovation (FIT), managed by the minister of economic development, was oriented toward industrial research and pre-competitive development. It was focused on combining the system of research support with the needs of the Italian industrial system. For the implementation of these measures, the granting of contributions was entrusted to one or more asset management companies (SGR *Società di Gestione Risparmi*) identified by tendering procedures from the minister. Moreover, the limitations for investments for big enterprises (as provided by the first FIT) have been lifted (MISE 2010). Differently, the Fund for Applied Research (FAR) was managed by the MIUR, and it was mainly oriented toward industrial research, including training and dissemination of technologies deriving from the activities financed by the former. It aimed at stimulating industrial research projects autonomously presented by companies, and at favoring the establishment of new companies (MISE 2010, 34). This fund represented a fundamental innovation in the national R&I instrumentation because, for the first time, industrial research was funded with a strong focus on public-private collaborations, mobility of researchers between industry and public research, and spin-off activities (Pierantozzi 2007). The underlining logic of this reform was to facilitate professors, researchers, and their institutions wanting to transform a valid research project into a business venture of potential success, through the creation of spin-off enterprises. It was inspired by similar instruments introduced in the French system (the SAIC, Industrial and Commercial Business Services, which was actually better targeted to this objective), with the goal of proving the public research sector with the opportunity to pursue the creation of new business initiatives (Cobis 2008).

The FIT and the FAR funds were meant to integrate each other's actions, despite their long-term tendency to overlap in their actions. This issue was primarily related to the overtime reduction of the financial allocation for both funds, the general reduction in the number of applications, as well as the layering of multiple policy interventions upon their action (MISE 2010, 21–23). As we will discuss in the following paragraphs, these funds are representative instruments for the Italian R&I system. These have been active for the whole period under investigation, despite the discontinuous allocation of resources which translated into a problematic implementation that negatively affected policy recipients' take up.

Finally, as a final component of this reform, a national development agency, called *Sviluppo Italia* (Development Italy) was established in 2000. This organization had the mission of promoting regional development and attracting investments with the underlining goal of supporting the development of new technologies in the country. The agency had the task of supporting professors, researchers, and students interested in developing and marketing the results of their research activities.

The *Bassanini reforms* focused on improving the efficiency of the national R&I system by encouraging managerial and administrative decentralization. Indeed, by means of a less coercive instrument mix, and the creation of an intermediary agency, it reduced the distance between policymakers and instruments' recipients. This reform mainly addressed the coordination, planning, and evaluation of the national research policy but it did not introduce any strong link between the resources made available and the thematic areas of the national research plan (European Commission 2006). In parallel, the country was undergoing a wider devolution process, which awarded regions with increased autonomy in the design and management choices for R&I policies. The national government preserved only the necessary competences to monitor and control their implementation in order to transfer best practices around the national territory (European Commission 2002). However, the titularity of the central government in this field was tight with the necessary approval that regional innovation plans needed to obtain from the minister of economy and finance, the minister of higher education and research, and the minister for economic development, to make sure the promoted initiatives were in line with the national R&D guidelines and policies. The financial resources involved to support such programs usually tended to be shared between regional and national governments and drew upon European structural funds.

5.2.1 *Plan for Digital Innovation in Enterprises*

In July 2003 the Ministry for Innovation and Technology (MIT) and the Ministry for Economic Development launched the action plan for digital innovation in enterprises, which implemented an integrated framework of measures promoting ICT-based innovations in the national economy. The plan supported SMEs' access to information technologies and technology transfer activities (European Commission 2003). It was implemented throughout two different waves (2003–2005), both involving the cooperation of different ministers with R&I competences, under the leading role of the minister for economic development and the minister for innovation and technology, in close collaboration with regional actors. The overall policy mix revolved around a combination of incentives, regulations, and measures devoted to coordinating public and private investments in technological innovations. The instrumentation was highly integrated not only across the two plans but also within existing policy mixes, as demonstrated by the thematic calls for innovation projects. Both digital innovation plans relied on expenditure instruments to support SMEs' projects, as well as on information instruments (consulting services) to support interactions with the public research sector. For what concerns technology transfer initiatives, the policy mix relied on information-type instruments, directed at improving the competitiveness of the productive system by strengthening and integrating the available supply of services for R&I activities. Some instruments have been specially re-financed after the two plans because of their important contribution to the promotion of the competitiveness in the productive system (Gallo et al. 2008). Despite the predominance of low automatic type of instruments, the mix appeared to increasingly rely on the collaboration of R&I performers and on the more automatic type of delivery structures. Another important innovation of the plan was the joint policymaking efforts of the two ministers, especially for creating specific collaboration between other ministers and regional actors. Indeed, the design of the two plans relied on a soft administrative decisional structure between the Ministry of Innovation and Technology and the Ministry of Economic Development, to identify the main lines of interventions. These were then discussed with the involved sectorial ministers and regions, with the open possibility to consult R&I performers.

5.2.2 *The Strategy "Industria 2015"*

In the period between 2000 and 2006, R&I strategies were almost absent from the governmental policy agenda. The political interest in this sector mainly materialized through one-off interventions included in the yearly budget, except for a reform influencing the governance of public research institutions and universities.

With the election of the new governing coalition lead by Romano Prodi (2006), policymakers engaged in the promotion of R&I activities in the country, with the implementation of the "*Industria 2015*" strategy. This plan aimed at ensuring a strategic repositioning of the national economic system by creating an environment favorable to innovation and technology transfer (Traù 2009). It defined major country goals on which to converge public action and the commitment of those actors operating in the market. The program designed, for the first time, a medium-long-term national strategy, with the goal of overcoming the dominant legacy of one-off instruments. A strategy that has proven to be inefficient, due to the instability of the funding allocations, as well as the wide discretionary space left to bureaucracy in funding allocation procedures (Sirilli 2010).

Differently, Industria 2015 involved a high degree of governmental planning resources. This resulted in an instrument mix including a combination of both automatic and generalized instruments, together with more selective measures tailored to the specific needs of recipients (as approved through the budgetary law for 2007; l. 296/06). The strategy revolved around: information and expenditure instruments supporting industrial innovation projects (IIPs); expenditure instruments helping (through public guarantee) enterprises' access to finance and automatic instruments; fiscal incentives supporting the implementation of innovative strategies by enterprises; and legal provisions facilitating the creation of business networks among SMEs.

The first pillar of the strategy was the competitive and development fund. This instrument financed industrial innovation projects, public innovation actions aimed at coordinating activities of large-scale public and private enterprises, industrial, and technological districts and the world of research and innovation. Their objective was to encourage the creation of partnerships between universities, research centers, private enterprises, and financial capital. Each project was led by a private manager, selected by policymakers based on their experience as private sector manager in different innovative fields. These profiles coordinated the interests of the innovation actors involved in the innovation project while overseeing the identification of the right instruments to be adopted for the implementation of the innovation project. This strategy was focused on five areas of intervention identified by the government (energy efficiency, sustainable mobility, new technologies of life, new technologies for Made in Italy, and innovative technologies for cultural heritage), and it gathered national policymakers, local authorities, industrial actors, universities, public research organizations, and the financial actors[8] to identify innovative projects for national technological and productive areas. The minister for economic development was coordinating the application and the funding system, through which industrial and research actors were invited to

[8] http://www.industria2015.ipi.it/index.php?id=2

create consortia and submit innovation projects. These proposals were then examined by the National Agency for the Diffusion of Technologies for Innovation, an institution created by the 2006 budgetary law (Berlusconi II government), according to technical-scientific and economic criteria.

Industria 2015 brought an important innovation in the national R&I policy design. With this strategy, public support for industrial innovation process overcame the correspondence between incentive-specific activity, and it assembled instrument mixes according to the initiatives to be realized. Industrial innovation projects were the result of strategic agreement and cooperation between the Ministry for Economic Development, the Ministry of University and Research, together with the Ministry for Innovation in Public Administration. Indeed, these three policymaking institutions, under the direction of the minister for economic development, coordinated their economic and managerial competences along the implementation of the *Industria* 2015 strategy. Such a degree of institutional coordination was critical during the selection of private project managers and of industrial innovation projects. This collaborative approach was also reflected by the high share of mixed and automatic delivery structures adopted.

Moreover, *Industria* 2015 reorganized existent funding mechanisms, not only in terms of policy rationality but also from a technical perspective, by combining new instruments with already active tools. It reorganized the fragmented landscape of existing measures by combining different funding structures into the competitive development fund. Moreover, it designed the FIRST (Scientific and Technological Research Investment Fund), which was supposed to gather different funds under the titularity of the minister of university and research: the FIRB (mission oriented and basic research), the FAR (fund for applied research), and the PRIN (fund for basic research). The funding rationalization process was also supported by the creation of an organizational structure based on IIPs, which helped to concentrate resources on large projects and specific key sectors (Potì and Reale 2010). Within this system, national policymakers, together with local authorities, industrial actors, universities, public research organizations, and the financial system[9] identified technological and productive areas of interventions as well as related innovation goals to be achieved. Despite the large room for maneuver left to project managers in shaping

[9] http://www.industria2015.ipi.it/index.php?id=2

the characteristics of each industrial innovation project, *Industria* 2015 represented the first programmatic governmental effort to steer R&I policies toward specific policy goals (Silva 2007). The underlining (soft) mission-oriented purpose of this strategy was modified by the following elected cabinet, led by Silvio Berlusconi, which granted the minister for economic development the power to add further technological areas to the six sectorial missions identified by the first policy (budgetary law for 2009 l. 133/2008) (Traù 2009). Together with the introduction of new instruments across different legislative interventions, as described below, it basically changed the mission-oriented nature of the strategy.

5.3 The Reforms of Public Research and Higher Education Policies

The governance of the Italian public research system has historically been grounded on the leadership of the National Research Council (CNR). This organization used to play a double role in the national research landscape, being both a research funder and a research performer.

Indeed, the council oversaw the administration and the selection of national research projects, and it allocated grants, so-called finalized research projects, to individual researchers. At the same time, the council was also undertaking its own research activities funded through mission-oriented research grants earmarked by funds from the Interministerial Committee for Economic Planning (CIPE). Therefore, on the one hand, the research council was acting as a funding agency, allocating competitive research grants to individual researchers while, on the other hand, it was also undertaking its own thematic research agenda.

With the establishment of the Ministry of University and Research in 1989, and the implementation of the Bassanini reform, the governance of the sector was transformed. Responsibilities for funding allocations were moved to the newly established minister, and the landscape of public research organization was also reorganized.

In this new setting, the national research council continued to benefit from mission-oriented research funding for its own research activities, the "strategic research projects" (Sirilli 2010), but its role of funding agency was terminated. However, the relevance of this organization was slowly transformed by hollowing out its funding competences. Moreover, the allocated funding for research (strategic research projects) was concluded in the framework of an overall reform of public research organizations, implemented in 2003. This set of reforms was a paradigmatic policy change in the governance of the national research sector since it transformed a quasi-research funding agency into a research performer. Basically, finalized research projects began to receive less funding to be ultimately replaced by two new instruments, directly managed by the minister of university and research: the fund for basic research investments (FIRB), and the special applied research fund (FISR). With this new instrumentation, the minister gathered competences in funding allocation and the authority to select areas for thematic funding. These evolutions ran in parallel with a reform process in the higher education sector. Changes in the recruitment system and in the internal governance of universities were the principal transformations in the sector. In line with other higher education reforms in Europe, policymakers transformed Italian universities into organizations inspired by a managerial-oriented logic, with a specific focus on research and teaching performance for securing public funding allocations. However, despite these two major reforms, the governance of Italian higher education, and public research, has been historically characterized by a dynamic of push and pull between the State and universities. Policymakers attempted to implement different rationalization strategies in the sector, while universities tried to resist in maintaining their traditional model refuting the push toward increasing autonomy embedded in many reforms (Capano et al. 2016; Reale and Potì 2009).

5.3.1 *The Reforms of Higher Education Institutions*

The beginnings of the Italian higher education system as policy sector can be dated back to 1989, when the Ministry for University and Research was established (l. 168/89). This was the result of the merger of competences on scientific research and university education that had previously pertained to the Ministry without Portfolio for the Coordination of Scientific and Technological Research, the Office of the prime minister, and the Ministry of Public Education (Capano et al. 2016). Before this reform, the governance of the sector was in the hands of different university-related powerful oligarchies and negotiations among ministerial bureaucracies (Capano 2008). In the reformed governance, universities were allowed to establish their statutes without any constraints from ministers. These organizations were granted with greater organizational and scientific freedom in defining their own regulation, management, and financial strategies. However, due to the low rate of implementation of university internal status, in 1995 the government approved a regulation forcing universities to do so, and within two years all Italian universities did manage to follow the new regulation. This case of "imposed autonomy" is symptomatic of the obstacles universities were experiencing in making autonomy happen (Capano 2008).

Until 1994, the system for funding state universities was dependent upon governmental decisions, and it came together with restrictions on how these funds should be spent and distributed within the internal governance of universities. This system was reformed with the introduction of the Ordinary Financing Fund (FFO), a lump-sum transfer assigned annually through budgetary law and devoted to cover universities' operating costs. This instrument contained two funding rationales: a share of funds allocated according to an "historical principle," hence according to the amount received by the institution before the reform; and a "balance share," allocated according to criteria defined by the minister for university and research. The "balance share" was replaced in 2004 by a "formula funding" introduced by the Committee for the Evaluation of the University System (CNVSU) based on the assessment of teaching quality and on the performance of higher education institutions. The "historical principle" funding model was then combined with a share of funding tied to a performance evaluation procedure. This established that one-third of the funds were assigned according to the number of students enrolled, one-third according to the outcomes of the educational process, and the last share according to the number of researchers employed, their rate of success in securing national research funding, and the income resulting from research commissioned from outside the university. This funding instrument combined resources necessary for supporting university-standard operating costs, together with funds allocated according to a reward nature. Therefore, such a combination of funding rationales highlights how policymakers purposefully decided not to distinguish between different policy goals and the instruments' features necessary for their achievement (Banfi and Viesti 2016).

With the approval of a reform of the higher education system in 2010 ("*Gelmini Reform*"—named after the minister who designed the policy), public support for universities was further transformed. This policy implemented different reforms in the governance of the system, and it emerged in a general climate where universities were perceived as "enjoying autonomy, without responsibility." Moreover, pressures from similar reforms also applied to other European countries (e.g., the Law on the Liberties and Responsibilities of Universities in France) and were also leveraged on in the Italian context to legitimize the implementation of the reform.

First, the reform modified the internal governance structure of Italian universities. It changed the system of power distribution between internal bodies, increasing the verticalization of power and decision-making within universities' managerial structure (Capano et al. 2016). University rectors were assigned with greater powers to design strategies for their institutions, as well as with the titularity to nominate general directors (responsible for university technical and administrative staff). Both the academic senate and the board of directors were changed in their internal composition granting the former the power to issue proposals in different areas (e.g., budget, research, teaching) and to design internal regulations. The board was granted with the power to orient institutional strategies together with decisional powers with respect to budget, financial planning, administration, and accounting functions. Moreover, a certain number of the board of directors could now be individuals not directly involved with university administration (for at least three years). However, the procedure for their election was left to the autonomy of universities' statute choices.

Second, the reform revised national provisions for the recruitment of researchers and professors. In line with the push toward the autonomy of the university, it provided departments with the powers to choose their organizational research profile as well as with the freedom to determine their recruitment policies. Moreover, it introduced a mandatory obligation to pass the National Scientific Qualification (NSQ) test for academic personnel. This procedure assessed individual eligibility for academic positions, relying on quantitative, bibliometrics scores, and a qualitative peer-review of the candidate's curricula and criteria. An evaluation committee, composed of five full professors appointed by the Ministry of University and Research, certifies the scientific qualifications necessary for the application to run for positions of associate or full professor. Once acquired, this certification lasts for nine years, but it does not grant access to any tenure position. It simply assigns a "label" to researchers allowed to run for these positions, whose recruitment dynamics are left to the management of universities' departments.

Quantitative metrics for assessing candidates were produced by a public agency, the Italian National Agency for the Evaluation of the University and Research Systems (ANVUR) (Bologna et al. 2021). The evaluation agency was officially created in 2006 (d.lgs. 262/2006), under the left-wing Prodi II cabinet. However, it only became operational in 2011, after experiencing a change in government composition and a shift in the political majority. It was responsible for the external evaluation of the activities of universities and research organizations. The agency had the task of evaluating national research performance, assessing the quality of university education programs, as well as estimating doctorates and universities' third mission strategies. With the establishment of an evaluation agency for universities and research institutions, policymakers introduced greater emphasis on performance indicators as a criterion for funding allocation. More generally, after the implementation of the *Gelmini reform*, concepts like performance (teaching and research), and indicator-based evaluation, became relevant factors in the allocation of public funds. Indeed, each university introduced an Internal Evaluation Committee (*Nucleo di Valutazione*), with the task of assessing teaching, research, and service-based activities implemented within each institution.[10]

Performance-based indicators produced by the activities of the National Evaluation Agency also influenced the allocation of the Ordinary Financing Fund especially in the calculation of its "balance share," which was now distributed according to different indicators produced by the ministry in accordance with the results achieved in the research evaluation exercise and in the evaluation of recruitment polices. This rationale for the distribution of funds was deepening the pre-existing national divide characterizing the performance of public research performers. Indeed, this logic tended to favor those higher education institutions already capable of dealing with increasing student fees and attracting research funds, usually localized in the wealthier areas of the country while also intensifying the geographical gap between the institutions left behind (Banfi and Viesti 2016; Capano et al. 2016). Indeed, the more historically productive, or prestigious, universities would manage to obtain more funding, while the worst-off would see their position further deteriorate because of the inaccessibility to new funding.

[10] An important new duty of this body was to connect institutional governance with system governance: besides submitting the results of its assessment to the senior management of the university, which could use them for strategic planning, it also acted as an operating arm of the ANVUR within each university (Capano, Regini, and Turri 2016, 107).

5.3.2 The Evolution of National Public Research Organizations

The governance of public research has historically been revolving around two different focal points. On the one hand, a minister (without portfolio) under the presidency of the council of ministers for the coordination of scientific research while, on the other hand, a National Research Council, which had both the role of public research performer and funder.

The *Bassanini* reform intervened in the sector by altering the system supporting industrial, public research, and experimental development led by the minister of university and research. The structure of funding for public research was reorganized (ministerial decree 593/00), with the aim of re-centralizing the responsibility for project funding allocation in the hands of the ministry, together with a reorganization of existing public research institutions and instruments. Consequently, the role of the National Research Council as a funding agency was interrupted. The finalized projects managed by this organization started to receive lower funding provisions, ultimately to be terminated. Consequently, the National Research Council transformed its nature, from being both funding administrator and recipient toward the sole role of recipient, as every other public research institution in the country. Other instruments funding basic research were managed by the minister of university and research, as the Research Programs of National Interests (PRIN),[11] an annual blue-sky research fund targeted toward higher education institutions. Differently from the projects managed by the National Research Council, the PRIN included a system based on experts' assessment, which evolved into a formally structured peer-review examination (Potì and Reale 2007). After the approval of the budgetary law of 2001, the funding landscape was supplemented by the Fund for Basic Research Investments (FIRB). A competitive funding instrument supported fundamental research objectives identified in the National Research Plan.

After some structural reforms in the sector, the national policy mix supporting research followed a discontinuous evolution pattern, characterized by specific one-off interventions approved in the framework of various budgetary laws. Two important interventions characterized the 2000s. The first one concerned a change in patent regulation, approved in the framework of the budgetary law for 2002, while the second is a tax

[11] In ministerial documents the acronym PRIN has been differently adopted, despite the reference to the same period of funding instruments. Until 1997 it was called 40% PRIN, from 1997 to 2004 COFIN-PRIN, and from 2005 PRIN (Corradi 2009, 147).

incentive addressed to individual researchers, approved with the budgetary law of 2004. With the modification of the patent regulation system, the so-called professor privilege (Lissoni and Montobbio 2015) was introduced. Therefore, despite professors having always enjoyed a free hand in dealing with IPR matters (Lissoni et al. 2009, 601), after this new regulation researchers became the owners of the rights deriving from the patentable inventions produced as results of their research. Universities, and public research centers, now have the authority to establish the maximum amount of the license fees for the use of the invention by third parties. Nevertheless, in any case, the inventor was entitled to a minimum of 50% of the proceeds, or fee, for the utilization of the invention. While the second instrument, adopted by the Berlusconi II cabinet, promoted a tax reduction for non-resident researchers—either Italians working abroad and wishing to come back to their home country or foreign nationals residing abroad and wishing to work in Italy. Its goal was not only to combat the widespread "brain drain" phenomenon tackling the Italian knowledge sector but also to support the technological and scientific growth of the country by easing the inward movement of new researchers.

The diachronic evolution of policy mix in the sector of public research reflects the instrument polarization of the national R&I system and the "residual" approach adopted for universities and public research organizations, especially after the beginning of the 2000s. This trend is also evident in comparison with the heterogeneity, and especially the higher share of instruments devoted to business actors, as discussed in the following paragraphs.

5.4 The Evolution of Industrial Actors

The Italian economy, similar to that of France, has developed by means of state participation. Specifically, it relied on public institutions like the Institute for Industrial Reconstruction (IRI—*Istituto per la Ricostruzione Industriale*) and the National Hydrocarbons Board (ENI—*Ente Nazionale Idrocarburi*). These stakeholders formed part of the state holding enterprises, which administered the majority of shares in national leading companies (Spence 2014, 105). However, at the beginning of the 1990s, the country undertook a process of disposal of state participation in the processing industry, both in the banking system and in the utilities sector. This was in line with the privatization wave characterizing many Western European countries, as well as in connection with the political failure in managing national holding models of industry. Such privatization has initiated the exit of public actors from those stakeholders potentially able to provide directions to the industrial development of the country (Traù 2009).

During the decade between 1980 and 1990, national policy design for industrial R&I was primarily financing projects submitted by firms, while automatic incentives were mainly directed toward the purchase of machinery. The sector was characterized by a political void, policymakers were reluctant in identifying sectorial missions for industrial development and technological innovation, in discontinuity with the dissolving era of state-run-enterprises. Hence industrial actors were presenting, in a bottom-up manner, their innovation projects to policymakers, which had now the task to assess their value and development potential. This system endogenously favored only those industrial actors with the economic capacity and technical resources to propose and implement innovative projects. Meanwhile, it cut off industrial actors, like SMEs, that often struggle (for reasons related to size and industrial specialization, among others) in developing autonomous innovation strategies.

However, SMEs represented a solid national productive structure employing medium- to low-skilled employees (European Commission 2004), which could benefit from policy interventions supporting innovation. Therefore, the innovation instrumentation implemented at the end of the twentieth century was exactly focused on solving these system imbalances. It consisted of actions directed at supporting the innovative efforts of SMEs, namely:

- a tax incentive for R&D-related expenses (l.140/97)
- targeted funding, implemented through the Treu Law (l.196/97), supporting the costs of full-time employment for professionals with PhDs in research activities undertaken by SMEs
- a tax credit (l.449/97) focused on alleviating the costs of hiring research personnel coming from the public research sector for a maximum of four years (after which they could come back to their institution of origin)

This instrumentation combined the use of fiscal incentives, and expenditure instruments, by SMEs to support R&D projects carried out on their behalf by public research laboratories or through the employment of professionals with PhDs. The underlying goal of this policy mix was to establish more permanent collaborations between SME's stakeholders and public research institutions. In the framework of national administrative system reorganization discussed in previous paragraphs (Bassanini packages), some of these measures were delegated to regions (l.140/97), while also centralizing competences under the measures financed through the FIT and the FAR funds.

With the beginning of the new century and the election of prime minister Silvio Berlusconi, SMEs remained the leading ladies of industrial innovations. However, with the rupture of the newly elected governments, and especially due to the strong emphasis of the new cabinets on alleviating tax burdens, the instrumentation for SMEs was heavily re-oriented toward fiscal instruments. SMEs were now relieved from taxation on R&I-related activities, together with an increasing involvement of the private banking system in the implementation of these tools. However, despite a clear political mandate to alleviate fiscal burden, this policy strategy proved not to be entirely homogenous since it was implemented by means of one-off interventions included in annual budgetary laws. The "*Tremonti Bis*"[12] is the first example of one-off intervention policy design logic. It consisted of a tax reduction oriented at sustaining the re-modernization of existing (or the construction of new) industrial plants and machinery, the completion of pending works, and the acquisition of new tools/instruments and items (also leasing contracts). Inspired by analogous measures approved in 1994, it was improved in its action by also including expenses from training events for personnel (European Commission 2002). It only included innovation-based expenses performed during the following fiscal year.[13] This measure was followed by a very similar, and short-lived, instrument: the so-called *tecno-Tremonti* (in force only during 2004). This was again an automatic fiscal incentive, stimulating SMEs' R&I investments by reducing their taxable income through a combination of R&D expenses and increments. Here, for the first time, the list of eligible expenses for fiscal deductions was broadly defined; although, it resulted in an increased likelihood of evasive behaviors by recipients (on this point see Bersani et al. 2004, p. 16).

In parallel with the layering of one-off fiscal instruments from SMEs, a more substantial transformation of national instrumentation was taking

[12] Law 383/2001, article 4–5.

[13] Between July 1, 2001, and December 31, 2002.

place through the introduction of a mixed lending system open to the bank sector.

Ministerial expenses (e.g., funding for expenditures and fiscal instruments) were progressively transferred to the Italian Credit Securitization Company (*Società di Cartolarizzazione dei Crediti Italiani*). This allowed for transforming outstanding credits into short-term financing and supplemented the progressive decrease in the resources available for the Ministry of Higher Education and Research and the Ministry for Economic Development.

The Deposits and Loans Fund (*Cassa Depositi e Prestiti*, CDP), a national public administration body, was transformed into an investment bank operating on the behalf of the government (its main shareholder). Basically, the Deposits and Loans Fund allowed for the participation of private shareholders in its capitals,[14] hence becoming the financial intermediary to which were sold the shares of different companies held by the Ministry of Economy and Finance. This stakeholder became effectively operational after the implementation of the revolving fund for enterprise investments (FRI). This fund granted access to easy terms credits for companies, and it relied on the Deposits and Loans Fund as manager for loan implementation. At least 30% of the funding made available was directed toward activities supporting strategic R&D projects of companies to be carried out jointly with public research actors. Recipients could get subsidized loans for up to 50% of the required financing, under the following contractual conditions: a share of the financing (at subsidized rate) was provided by the Deposits and Loans Fund, and the other share by private banks at market rate. In both cases the economic judgment of the R&I plan submitted by the recipients was analyzed by the partner bank.

The introduction of an innovation credit financing system, based on banks, was the major novelty of the Berlusconi II cabinet. It unloaded ministers with duties related to the implementation of funding instruments, and it involved banks as intermediary private actors. This reform aimed at easing the planning of resources for investments in innovation by enterprises, but it didn't provide any specific policy goal in terms of national R&I strategy (Gallo and Silva 2006).

[14] In May 2015 the shares were divided as follows: 80.1% Italian Government, 18.4% Banking Foundations, and the remaining 1.5% in Deposits and Loans Fund internal treasury shares.

A new wave of reorganization in the structure of incentives for territorial and industrial development started in 2009, when start-up companies began to populate the Italian innovation system. This introduced instruments promoting R&D investments by firms, like the Italian investments fund and the JEREMIE fund (2010), equity-fund-based instrumentations reinforcing the role of private investors in start-up innovation finance. The former supported the development of SMEs' innovation projects through private equity financed by private banks and employer organizations.[15] Meanwhile, the latter, the JEREMIE (Joint European Resources for Micro to Medium Enterprises Fund) fund, was an equity fund for temporary and minority participation in SMEs supporting the implementation of innovation projects in the framework of European structural funds (Potì and Reale 2011). The development of innovative start-up companies was further enhanced through the creation of the National Innovation Fund (2011), supporting SMEs in accessing risk capitals. This instrument relied on the cooperation of private banks involved in the evaluation and project funding phases. These funding instruments were further supported in their development through a new legislation defining the characteristics of venture capital funds.

In conjunction with the Berlusconi IV cabinet, Italy was caught in economic and financial troubles due to the worsening of the sovereign debt crisis in Europe and the financial turbulence triggered by the 2008 financial crisis. The government resigned and a technical government, led by Mario Monti, was appointed by the President of the Republic. This cabinet was operating under strict budgetary constraints, mainly aimed at the financial stabilization of the country. It mainly focused on the simplification of the funding opportunities available to stakeholders in the R&I system (Nascia et al. 2012) and on supporting the employment of highly skilled workers in, potentially, high innovation intensity sectors.[16]

[15] Banks such as Intesa San Paolo (Bank), Unicredit (Bank), Nexi (bank), and the Trade Association of Italian Banks; employer organization: the General Confederation of Italian Industry (Confindustria).

[16] The Council of Ministries commissioned to an external expert, Francesco Giavazzi (an economist, professor at an Italian university), the task of reviewing the Italian system of government subsidies to firms. This report estimated a total amount of "unjustified" subsidies and argued for their elimination with a parallel tax cut (Lucchese et al. 2016, 248).

5.4.1 *Industrial Districts and the Complexity in Promoting Policy Change*

The Italian industrial system is largely made up of SMEs which tend to be mainly family-run businesses (Pierantozzi 2007). Their development, and proliferation, revolved around a specific organization of the productive system, characterized by geographical agglomerations of enterprises specialized in a related sector of activity: the industrial district. These are homogeneous production systems with a concentration of (small and medium-sized) industrial companies, with high productive specialization, where enterprises integrate their activities in the same production process. Districts have historically emerged through the autonomous concentration of geographically close industrial actors, especially in the field of textiles, clothing, mechanics, and leather.

The first legal recognition of industrial districts dates to 1991 (l.317/91). According to the guidelines identified by the Ministry of Finance, each region oversaw the identification and funding of enterprises. Only in 1999, with the so-called *Guarino decree* (l. 140/99), policymakers defined the distinctive criteria characterizing industrial districts. The identification of regional industrial districts was basically a bottom-up process, involving regions and local actors while the minister for economic development defined the standards for their identification. This effort was accompanied by greater empowerment and autonomy for regions in supporting the development of industrial districts, due to their geographical proximity.

Industrial districts were supported through a mix of ordinary national budgets and resources from regional policies (which were themselves a mix of European structural funds and budget allocation from thematic funds).[17] The instrument mix supporting their development was composed of administrative simplifications for accessing fiscal instruments and

[17] The Berlusconi II cabinet introduced, with the budgetary law for 2003 (l. 289/02), a fund for underutilized areas, called the FAS. This fund, together with structural funds for national and regional operative programs (PON and POR), financed regional policies for economic development.

tax benefits.[18] Industrial districts have been the backbone of the Italian industrial sector and had often been portrayed as the showpiece of national industrial capacity, no matter the political orientation of the cabinet in office.

However, the system started to show some weaknesses primarily related to the industrial and economic structure of involved SMEs. Italian SMEs tend to be characterized by a low degree of internationalization, low investments in innovation activities, the lack of consolidated management culture, as well as the inability to scale up production (M. Di Maio 2014). Therefore, to keep up with industrial and technological development, a new format of industrial districts was implemented (in 2002) by the Ministry of Higher Education and Research: the technological districts. This intervention shared many similarities with the "original" industrial districts. They were both related to the geographical proximity of the actors involved, as well as to the benefits connected with the underlining dynamics of their interaction. Nevertheless, technological districts differentiated themselves for two reasons. First, they were not limited to SMEs; also, large and technologically intensive firms were involved. Second, they were mainly working with highly skilled human resources and local public research centers.

Technological districts were meant to facilitate cooperation between scientific and technological players together with companies, and to support joint collaborations for developing competitive research projects (European Commission 2004). Their goal was to establish R&I networks, territorially embedded on specific technologies, with the collaboration of small and large firms and with a strong orientation to the socio-economic valorization of results (Poti et al. 2008, 33). Despite their close similarities with industrial districts, their design was presented as an innovative strategy to support the Italian competitive capacity internationally (Ministero dell'Istruzione dell'Università e della Ricerca 2005). From a policy instrument perspective, technological districts were a public-private-partnership

[18] Budgetary law for 2006 (l.266/2005) the Berlusconi cabinet II outlined the characteristics and the methods for identifying industrial districts and it also defined their tax-related regulation. Budgetary law for 2007 (l.296/2006) and supplementary decree (l.112/2008)—the Prodi government simplified the tax regulation for industrial districts, these provisions were modified again by the following Berlusconi III cabinet (d.l. 5/2009) and were supplemented by the introduction of further tax benefits for enterprises involved in industrial districts (l.122/2010). Finally, the Monti cabinet simplified the procedures required to establish industrial districts (d.l. 83/2012).

between local public research and industrial actors. Regional actors were the institutional promoters of these initiatives; they created the coordination table at which local interested actors could participate, and they presented the official proposal to the Ministry of Higher Education and Research which had power over funding decisions (Cobis 2008). Moreover, technological districts have also been provided with their own funding stream, becoming a top priority in the reformed objectives of the fund to sustain enterprise investment (FRI).[19]

5.5 The Configurations of Ministerial Organization Strategies

The national governance of Italian R&I revolves around an external ministerial specialization strategy. Hence, R&I policymaking competences are decoupled among functionally homogeneous organizations: the Ministry for Higher Education and Research and the Ministry for Economic Development.

Historically, decisions concerning industrial strategies were taken by two different ministries: the Ministry of Industry (established in 1948) and the Ministry of State-holdings (established in 1956) (M. Di Maio 2014, 248). After the suppression of the latter (1993), the organization of the ministry with competences in industry has been evolving across different cabinet periods. An important step toward the establishment of a policymaking institution with competences in industrial policies was the *Bassanini reform*, which reunited competences on industry, national, and international trade into the Ministry of Productive Activities. This institution has been exposed to a tension between reunification and fragmentation across different cabinet periods. However, the institutional ministerial setting analyzed in this study (the Ministry for Economic Development) has been in force since 2006, and its internal organization into departments and directorates embeds the different competences that have been merged into one organization. Similarly, the Ministry of Higher Education and Research was created in 1989, merging competences on scientific research and university education that previously belonged to the Ministry without Portfolio for the Coordination of Scientific and Technological Research, the Office of the prime minister, and the Ministry of Public Education (Capano et al. 2016). This institution slowly acquired

[19] Reference (d.l. 35/2008, c. 4).

increasing powers in national research funding, in parallel with the decrease of the National Research Council's powers. The ministry experienced different organizational settings. However, the main reforms in the sector happened in periods of ministerial reunification in one institution of higher education and research competences.

The external ministerial organizational layout in R&I dates to the *Bassanini reform*, which reiterated the dual system of the division of competences according to the different titularity of funding allocations: the fund for applied research (FAR) and the fund for technological innovation (FIT). This organization of policymaking is supported by a linear understanding of the innovation process: the Ministry for Higher Education and Research deals with fundamental research, while the Ministry for Economic Development works on experimental development and market applications.[20] This underlines a policy rationality supporting the idea that a type of research can contribute to economic development, while another type cannot: an interpretation which doesn't always reflect how the innovation processes work (INTERVIEWEE 11).

National ministerial coordination strategies, which are defined as informal institutions in Chap. 3, suffer from poor implementation, as confirmed by most interviewees. The division of competences between ministries is blurred; this causes fragmentation and creates uncertainties in the design of national policy strategies. This negatively affects the capacity of the target population to navigate the R&I instrumentations at their disposal (INTERVIEWEE 10). Over the decades, policymakers have attempted to overcome these issues through the implementation of different, coordinated policy strategies.

The first was the creation of a National Research Program (NRP). This document identifies national strategies for R&I policies and their budget allocations every three years. The minister for higher education and research coordinates consultations with national R&I stakeholders, representatives of the business world, and all the public administrations concerned with the plan. Its implementation is conditional on the approval of the interministerial committee of economic planning (Lai 2016).

[20] The Ministry for Higher Education and Research covers the sphere of competences under the remit of the Frascati Manual, focusing on research activities that are slightly behind the market. Meanwhile, the Ministry for Economic Development referred to the definition of the Oslo manual and it was more concerned with how to transfer this research to the market—as confirmed by our interviewees (INTERVIEWEES 9–5).

Therefore, the position of the ministry in charge ultimately appears to be a formal, rather than an effective, role in policymaking. Indeed, (INTERVIEWEE 9): *"[…] it is difficult to see the NRP as a coordinated and coherent process; the sensation is rather that of a set of interventions based on isolated issues. The lack of coordination is systemic because once a set of similar competences is established among several ministers, these administrations are of an equal level, and it becomes difficult for a ministry to be the coordinator of the others because their weight and institutional level is the same."*

A second institutional strategy to support interministerial coordination and to enhance collaboration among peer institutions was to further involve the Interministerial Committee for Economic Planning (CIPE) in national R&I policy design. Its major functions included evaluating the Economic and Financial Planning Document, defining the general economic strategy of the country, and the annual allocation of resources devoted to R&I (European Commission 2006). The committee consisted of the prime minister (president), the minister of economy and finance (vice-president), the vice-minister of economy and finance (secretary), and other ministers whose presence was necessary to deploy the industrial and economic policies (European Commission 2004, 12). In the framework of ministerial coordination strategies, it represented a "table of directors"—an external platform where ministers with R&I competences could discuss national R&I strategies, especially funding allocations, under the coordinating role of the prime minister and the minister of economy and finance. After the election of the Prodi II cabinet (2006), the technical-administrative structures supporting the committee, until then operating within the Ministry of Economy and Finance, were transferred to the office of the prime minister. A strategy praised by many actors to be moving in the right direction, of centralizing power as a solution to overcome the policy deadlock caused by "peer-management" of R&I policy design (INTERVIEWEES 10-7-9). However, the centralization of power was not accompanied by political leadership in policy design choices. Indeed, the way this coordination body functions is left to the internal dynamics taking place within the cabinet, according to the different power games defined by coalition governments. Therefore, when ministers sit at this table, they are already aware of the power imbalances and of their related funding allocations, as if little room for maneuver were available in policy design choices (INTERVIEWEES 11-3-5). The committee represented a formal coordination table among ministers because *"policies are designed within the government and my feeling is that CIPE is not the platform through which public policies are effectively designed*" (INTERVIEWEE 3).

Despite the structural shortage of institutional coordination practices, the minister for higher education and research and the minister for economic development have managed to overcome institutional friction and have designed policies in a coordinated manner. These events are primarily the result of the entrepreneurial activities of some ministers who have interpreted more correctly, compared to others, these issues (INTERVIEWEE 9). As in the case of the *Industria* 2015 strategy, the successful design of the intervention is a credit to the minister of economic development, who managed to rapidly exploit an open political window of opportunity (INTERVIEWEE 8). However, the strategy did not survive the shift in political majority and the election of a new cabinet, which is symptomatic of the volatility of the policy strategies relying on the solo-entrepreneurial role of a minister (INTERVIEWEE 10). Differently, the collaboration between the Ministry for Economic Development and the Ministry of Innovation and Technology in the creation of the plan for digital innovation in enterprises has been smoother and more enduring. However, it is important to highlight some specificities that could have positively influenced this strategy. First, the strategies were formulated across two cabinets with a similar political majority. Second, their instrumentation mainly relied on the re-calibration of existing instruments toward the promotion of new technologies for enterprises, rather than on a brand-new set of instruments.

This evidence suggests how national R&I policymaking coordination is left in the hands of specific institutions, or ministers, and no clear coordination practice is provided at the national level. Indeed, "*it is exactly here where the main issue of the Italian R&I governance system lies, because if everything can depend upon the strength, or not, of a single minister, it means that the architecture of the system is not solid and perhaps it should be reformed*" (INTERVIEWEE 8). This governance failure is also perceived by R&I performers who, looking at the organization of the public structure, have the impression that the two ministers do the same job (INTERVIEWEES 10–11). The boundaries of competences between different policymaking institutions are often fuzzy, and this creates confusion between instruments and goals (INTERVIEWEE 8). This further hampers the policy capacity to design long-term national strategies and to identify national political priories in R&I. That is why, often, the evolution of national R&I policy mix is made up of fragmented instruments which look like the result of day-to-day administrative management, rather than of a national strategy (INTERVIEWEE 10). The organization of competences between these institutions should avoid the risk of falling into the trap of creating structures in which the bureaucracy derives its *raison d'être* only from the administration of subsidies.

5.6 The Fragmentation of Interest Representation

The infrequent coordination between R&I ministers in Italy is described as one of the factors that inhibits the achievement of good results in policymaking practice. However, a similar endemic fragmentation is also present on the side of industrial and business actors (Lanza and Lavdas 2000).

The representation of the industrial world in Italy hinges upon the General Confederation of Italian Industry (Confindustria), an association with voluntary membership, representing companies, their values and interests, at the institutional level. Nowadays, this organization is facing the same challenges that many representative bodies were already facing in the Italian political system (INTERVIEWEES 8–11). There are two contrasting views on the underlying reasons for this. The first Confindustria didn't truly believe in innovation as a motor for Italian economic development (INTERVIEWEE 6). While the second claims that big enterprises, once more involved in Confindustria and usually representing the biggest contributors in terms of national private expenses for R&D, have started to behave more autonomously in the national R&I governance and in their innovation investment choices (INTERVIEWEES 4-7-9). Indeed, when ministers are designing R&I interventions, in addition to the dialogue with Confindustria, they also organize one-to-one meetings with representatives of big Italian enterprises (INTERVIEWEE 5). Therefore, there is a consolidated perception regarding the increasing incapacity of this confederation to represent the requests of big enterprises in the field of R&I, which has been somehow translated into a greater attention toward the needs of SMEs.

The private sector has had an ambiguous role in national R&I governance. On the one hand, Confindustria has highlighted the need to enhance the production of high-quality human resources in the S&T field. While, on the other hand, the number of graduates hired by firms is among the lowest in Europe (Poti et al. 2008, 19). This contradiction seems symptomatic of a phenomenon defined as "the crisis of innovation demand of Italian SMEs" (INTERVIEWEE 10). Indeed, national small and medium-sized enterprises were overwhelmed by the instrumentation at their disposal, and the role of confederation should have been to accompany these actors during the selection of the most suitable instrument. Indeed, *"The confederation has always been very busy in dealing with the offer (establishing a dialogue with policymakers), but very little attentive to the demand side"* (INTERVIEWEE 10).

Alongside the loss of representative power from Confindustria, and more generally from the representatives of the public research sector, there are other types of interests that seemed to have gained ground, like the presidents of regions. Regions have increased their contribution to the national innovation efforts, also supported by European Structural Funds (European Commission 2006; Sirilli 2010). They represent local lobbying interests related to national productive structures (e.g., industrial districts), that began to intervene in the politics of R&I policy design (INTERVIEWEE 9). Therefore, Confindustria started to develop territorial strategies to coordinate with local representatives of the confederation and to open a dialogue with regional institutions (INTERVIEWEE 10).

Also, actors of the public research sector suffer from a representation issue. During the late 1990s, following the creation of the Ministry for Higher Education and Research, universities experienced a phase of "imposed autonomy" (Capano 2008). Prior to the introduction of the Ordinary Financing Fund (FFO), universities had an extremely low level of autonomy, and only with this reform universities became responsible for their budget expenses. Then, in 2010, the second systemic reform in the sector (*Gelmini Reform*) implemented a new dirigiste structure where universities were free to define their own statues, although under the condition defined by a tight ministerial regulation (Capano et al. 2016; Donina et al. 2015). This tension between imposed autonomy and dirigisme characterizes a higher education system where governments have tried to pursue different policy strategies, while universities have always tried to resist transformations (Reale and Potì 2009). When we zoom in on the system interest representations for higher education institutions, these organizations smoothly align into a relatively homogenous interest group. However, these institutions do not represent a powerful actor in R&I policymaking. There are some important research centers in the country that can stamp their feet on specific R&I policy issues; however, these actors were not powerful enough to guide the design of a policy and to be a stable interlocutor for the government (INTERVIEWEES 8-4).

Similarly, the National Research Council was also deeply transformed. The combination between different reforms of the public research sector[21] and a more general shrinking of public funding for research transformed the council into a technological service-oriented actor. This was meant to compensate for the reduction of public funding through the provision of some specialized services like analysis and technical tests, technological services, quality services, and environmental services (Coccia and Rolfo 2010). Before this transition, the council *"dirigistically selected which research sectors to push forward and in which direction. It used to behave as a public body,"* and it was perceived as an extension of the university world (INTERVIEWEES 6–7).

Despite the contrasting views (also among our interviewees) regarding the titularity, and the effectiveness, of its action there emerges a generalized consensus regarding its pivotal role in public research governance. At the end of this process the National Research Council lost its role of quasi-public funding agency, centralized under the remit of the minister for higher education and research to finally become a public research performer in the national R&I system. However, it seems this reform was forgetting to replace the council with another body that could perform the same function (INTERVIEWEE 7). Indeed, the Italian R&I governance is lacking a unified political responsibility center to whom different stakeholders of the public, but also private, R&I sector can raise their claims and suggestions (INTERVIEWEES 7-9-10).

5.7 Conclusions

The Italian national R&I system is affected by a weak systematic approach, by the lack of assessment and foresight activities, and by weak planning capacity in elaborating strategic programs (European Commission 1999). More generally, the system presents many internal contradictions in the way it functions. The *Bassanini reform* substantially contributed to reorganizing the national R&I governance structure. Although preserving the existent ministerial layout of external specialization patterns, it provided the Ministry for Economic Development and the Ministry of Higher Education and Research, respectively, with a funding instrument. However, despite the fact these two funds were meant to complement

[21] Bassanini package and a second reform implemented by the minister of higher education and research, Letizia Moratti, under the Berlusconi II cabinet (d.lgs. 127/2003).

each other's tasks, in many occasions they ended up overlapping in their actions (European Commission 1999).

The national system of innovation presents a typical top-down culture for coordinating knowledge demands supported by instruments like the National Research Plan and multiannual strategies. However, the fragmentation and the duplication of R&I instruments do still persist (European Commission 2006).

The evolution of the national R&I instrumentation doesn't seem to follow any clearly identified long-term strategy, embedding a high degree of misinformation between those who allocated the funds and the beneficiaries (A. De Maio 2011). When policies are then implemented, the overall framework displays a clear lack of coherence, and, given the layering of multiple measures, the resulting objectives are not often carefully calibrated for their long-term effects (European Commission 2005). Despite some positive experiences in the design of long-term national R&I strategies, as described at the beginning of this chapter, national policymaking was formed by the scarcity of a "holistic" policy approach. The dominant rationality seemed to be that of approving annual budgetary laws in which, on a case-to-case basis, different one-shot provisions were included. This results in a national R&I policy mix highly polarized between enterprise-oriented innovation initiatives and a "residual" approach toward higher education institutions and public research organizations (Nascia et al. 2012).

The diachronic evolution of R&I instrument features suggests an increasing relevance of collaborative types of instruments, for example, those based on the creation of public-private partnerships (like those found in the *Industria 2015* strategy and in technological districts). This tendency was also supported by the adoption of mixed type delivery structures relying on the collaboration of different actors to ease policy instrument implementation and uptake by recipients (like the inclusion of banks in the financing system for innovation). However, the increasing attention toward establishing collaborative relationships with R&I performers was not always matched with a sufficient effort in supporting, and stimulating, their demands for innovation. Indeed, Italian R&I instrument mix displays a consistent lack of culture and awareness about the possibilities this policy lever can offer to their recipients (Potì and Reale 2011). The attempts to shift national instrumentations toward a mix increasingly based on fiscal instruments have not always been effective, due to the overlapping of different initiatives and the lack of an overall consistent plan for action, which has been further exacerbated by the irregularity of public financing (European Commission 2000). Moreover, these interventions have not always been matched with measures of an informative, and educational, nature (e.g., information campaigns, consultancy services) necessary to support the demand of instruments for innovative activities by enterprises.

Appendix

Encompassing Research and Innovation Strategies

Bassanini Reform

Table 5.1 Instrument mix—*Bassanini reform*

Instrument	*Content*	*Shapes Family Delivery Target*
Reorganization of the research system (d.lgs. 204/1998)	Introduction of the National Research Plan (*PNR Piano Nazionale di Ricerca*), which programmatically defined research activities to be promoted. The budget allocations defined in this document were included in the national economic planning document (*DPEF Documento Programmatico di Economia e Finanza*), yearly approved by the parliament	Administrative simplification [Regulation # medium coercive] [Low automatic] [HEIs + PROs+ Enterprises]
Special applied research fund (FISR *Fondo Speciale Ricerca Applicata*) (d.lgs. 204/1998)	Financed research projects on thematic areas according to the guidelines of PNR. Mostly collaborative projects at national level. It was jointly managed by the minister of finance and the MIUR (who had the titularity for the submission of financing proposals). It required the co-financing of the public administration in question (30% of the budget required)	Targeted funding [Expenditure # low coercive] [Low automatic] [HEIs+ PROs+ Enterprises]
One stop shop for enterprises (d.lgs. 123/98)	It was a consulting service managed by municipalities (or an aggregation of thereof) aimed at simplifying administrative procedures related with industrial activities, thanks to the merger of different functions into a unique organization. Moreover, it also provided an information advisory and corporate support for companies wishing to embark on an industrial activity	Consulting service [Information # medium coercive] [Low automatic] [SMEs]

(*continued*)

Table 5.1 (continued)

Instrument	*Content*	*Shapes Family Delivery Target*
Reorganization of public research institutions (d.lgs. 19/99; d. lgs.27/99; d.lgs. 36/99	Aimed at reducing the overall number of Public Research Organizations and reorganizing their internal departmental structure. In order: CNR, ASI, ENEA	Administrative simplification [Regulation #l medium coercive] [Low automatic] [PROs]
FAR: evaluation component	Managed by the MIUR, it was devoted to finance projects autonomously presented by performers, international projects, projects for the creation of research infrastructures, projects devoted to increase competitiveness, creation of new enterprises	Competitive funding [Expenditure #low coercive] [Low automatic] [HEIs+ PROs+ Enterprises]
FAR: negotiated component	Managed by the MIUR, it was devoted to finance research projects on themes identified by the minister	Targeted funding [Expenditure #low coercive] [Low automatic] [HEIs+ PROs+ Enterprises]
FAR: automatic component	It contributed to the financing of recruitment for research personnel and outsourcing of research tasks to universities or public research organizations	Tax exemption [Taxation #low coercive] [High automatic] [Enterprises]
FIT: subsidized loans	Managed by the MISE, it contributed to experimental development projects. Devoted to research programs below €3 million and it covered 50% of the project expenses for max eight years. The subsidized rate of financing was equal to 20% of the reference rate into force	Loan [Expenditure # high coercive] [Low automatic] [PROs+ Enterprises]

(*continued*)

Table 5.1 (continued)

Instrument	*Content*	*Shapes Family Delivery Target*
FIT: interest subsidies	Managed by the MISE, it contributed to experimental development projects. Devoted to research programs above €3 million and related to a bank loan at market rate devoted to the financial coverage of the project. The interest rate contribution was 50% of the eligible costs, with a maximum duration of eight years (on the original loan provided by the bank)	Targeted funding [Expenditure #low coercive] [Mixed automatic] [PROs+ Enterprises]
FIT: targeted funding	Managed by the MISE, it contributed to experimental development projects. It was a quota of the FIT devoted to finance up to 20% of the costs for projects related with the areas and themes identified by the minister (left up to the minister, a share or the total FIT allocation could be devoted to a "program of relevant interest for the technological and productive development of the country")	Targeted funding [Expenditure #l ow coercive] [Low automatic] [PROs+ Enterprises]
Sviluppo Italia S.p.A.	Public sector national development agency, under the financial control of the Ministry for Economy and Finance. It had the task of promoting productive activities and attracting investments; promoting entrepreneurial activities; supporting public administration in financial and project planning; consultancy on the management of national and European incentives, especially for southern regions and depressed areas	Advisory committee [Regulation #low coercive] [Low automatic] [HEIs+ PROs+ Enterprises]

Plan for Digital Innovation in Enterprises

Table 5.2 Instrument mix—plan for digital innovation in enterprises

Instrument	*Content*	*Shapes Family Delivery Target*
Thematic call for innovation (Plan 2003)	By relying on the resources of the Fund for Technological Innovation (FIT) and the Fund for Underutilized Areas (FAS[a]) the plan identified specific calls for proposals to finance innovation projects devoted to the implementation of ICT-related organizational systems within SMEs	Targeted funding [Expenditure # low coercive] [Low automatic] [SMEs]

(*continued*)

Table 5.2 (continued)

Instrument	*Content*	*Shapes Family Delivery Target*
Thematic research and innovation calls (Plan 2003)	Creation of a specific fund devoted to finance research projects of significant scientific value, also with regard to health safety and technological innovation, for the years 2003–2004	Targeted funding [Expenditure # low coercive] [Low automatic] [HEIs+ PROs]
Riditt project (Plan 2003–2005)	The Italian Network for the dissemination of innovation and technology transfer was managed by Institute for Industrial Promotion providing information and technical assistance services to the national R&I performers	Consulting service [Information # medium coercive] [Low automatic] [HEIs+ PROs+ Enterprises]
Guarantee fund for SME (Plan 2005)	By relying on the resources of the Guarantee Fund for SMEs (approved with the l 662/96 and into force from 2000) a special section devoted to digital technologies was created. It provided public guarantee to SMEs in order to finance innovative projects related with ICT development. It facilitated credit access to SMEs, reducing the risk level of the credit given by banks and encouraging the consortium of enterprises in order to undertake common projects	Economic guarantee [Expenditure # medium coercive] [Mixed automatic] [SMEs]
High-tech fund (Plan 2005)	Fund for public participation in risk capital of enterprises operating in high technology sectors (IT, electronics, nano-micro technologies, electro medical instruments). The participation addressed already established or to be established funds or may be implemented through a direct support to venture capital activities. It aimed at promoting the creation of innovative enterprises in the high technology sector and attracting traditional venture capital	Equity participation [Expenditure # high coercive] [High automatic] [SMEs]

(*continued*)

Table 5.2 (continued)

Instrument	*Content*	*Shapes Family Delivery Target*
Technological vouchers	A financial facilitation (voucher) was provided to companies for the following activities: assistance to patent registration, business evaluation, technical due diligence, and PhD scholarships. Through this voucher SMEs and individuals could benefit from these services supplied by research institutes and universities accredited by the minister of innovation and technology	Consulting service [Information # medium coercive] [Mixed automatic] [HEIs+ PROs+ SMEs]

[a]Introduced by the Berlusconi II cabinet through the budgetary law for 2003 (l.298/02), see paragraph 7.7 on technological and industrial districts

The Strategy "Industria 2015"

Table 5.3 Instrument mix—*Industria 2015*

Instrument	*Content*	*Shapes Family Delivery Target*
Competitive and development fund (FCS)[a]	It financed IIPs related to the *Industria 2015* plan It brought together the resources formerly allocated to the FAS (fund for underdeveloped areas) and the Single Fund for Incentives to enterprises. The resources were allocated by the CIPE to the MISE	Targeted funding [Expenditure #l ow coercive] [Low automatic] [HEIs+ PROs+ SMEs]
Industrial innovation projects (IIP)	These were projects developed in specific areas, as identified by the *Industria 2015* strategy (energy efficiency, sustainable mobility, new technologies of life, new technologies for the Made in Italy, innovative technologies for cultural heritage). They were managed by private managers, with the task to identify specific support mechanisms, to involve different research and industrial actors, and to determine the project development	PPPs [Information# high coercive] [Mixed automatic] [HEIs + PROs + Enterprises]

(*continued*)

Table 5.3 (continued)

Instrument	*Content*	*Shapes Family Delivery Target*
Fund for corporate finance	It facilitated access to credit and risk capital for SMEs, providing public guarantees It is meant to be cyclically financed by the interest of the beneficiary enterprises in addition to public funding. The criteria and priorities for the implementation of the operations, also with reference to the recipient company and the eligible expenses programs, were established by the MISE	Economic guarantee [Expenditure # medium coercive] [Low automatic] [SMEs]
Tax credit for research in enterprises	Tax credit devoted to investments in specific geographical areas and expenses related with machinery, software, and patents The tax credit was given in proportion to the amount of expenses in surplus to the depreciation calculated for the taxable period, and it was under the titularity of the MISE	Tax reduction [Fiscal # medium coercion] [High automatic] [Enterprises]
Fiscal incentives for R&D investments	Tax credit equal to 10% of the expenses devoted to activities of industrial research and pre-competitive development. The tax credit increased to 15% in cases where the research contracts were stipulated with universities or public research organizations The ceiling for maximum expenses is equal to €15 million With the budgetary law for 2008 (l.244/2007), these ceilings were slightly modified. Indeed, the tax allowance for firms performing research activities in collaboration with universities rose to 40% of the expenses and the maximum amount of expenses to €50 million	Tax reduction [Fiscal # medium coercion] [High automatic] [Enterprises]
Business networks	This legal provision was devoted to the simplification of the legislation related with the creation of business networks among enterprises. It was aimed at enhancing organizational coordination among companies, especially SMEs, which aimed to cooperate, in the development of specific projects or in applying for specific funding but did not want to merge	Contract [Regulation # low coercion] [Mixed automatic] [SMEs]

[a]It will continue to finance existing law in the same way until the implementation decrees of reorganization of the legislation of subsidies come into force, but it will also finance IIPs. The basic idea is to improve the efficiency and effectiveness of the system by reducing the number of incentive tools

The Reforms in the Public Research Sector

Table 5.4 Reform in public research sector I—instrument mix

Instrument	*Content*	*Shapes Family Delivery Target*
FIRB (fund for basic research investments)	Competitive funding managed by the MIUR and financed through the budgetary provisions included in the FAR. It financed projects to upgrade large public or public-private research infrastructures, basic research projects of high scientific or technological contents, strategic projects for the development of pervasive multi-sectorial technologies, the establishment and networking of highly qualified scientific centers	Targeted funding [Expenditure #l ow coercive] [Low automatic] [HEIs+ PROs]

Table 5.5 Reform in public research sector II—instrument mix

Instrument	*Content*	*Shapes Family Delivery Target*
Patent regulation (l. 383/2001 art. 7)	In cases of employment relationship between a university or public administration having among its institutional aims research purposes, the researcher is the exclusive owner of the rights deriving from the patentable inventions of which she is the author	IPR [Regulation# high coercion] [Low automatic] [HEIs + PROs]
Tax incentives to non-residential researchers (l. 326/2003 art. 3)	The income from employment or self-employment of researchers who (from the date of entry until enforcement of this decree or in one of the five subsequent calendar years) began to carry out their activity in Italy, and who consequently become fiscally resident in the State, were taxable only for 10%, for the purposes of direct taxes, and did not contribute to the formation of the value of net production of the regional tax on productive activities. The incentive was applied in the tax period in which the researcher became fiscally resident in the territory of the State and in the two subsequent tax periods	Tax reduction [Fiscal # medium coercion] [High automatic] [Researchers]

Business Actors: The Industrial and Technological Sector

Table 5.6 Business-oriented I—instrument mix

Instrument	*Content*	*Shapes Family Delivery Target*
Employment in the field of research (l. 196/97)	It focused on the employment of professionals with a PhD to be hired for research activities in SMEs	Targeted funding [Expenditure #low coercive] [Low automatic] [SMEs]
Employment of researchers by SMEs (l. 449/97)	It focused on the employment of professionals with a PhD to be hired for research activities in SMEs	Tax reduction [Fiscal # medium coercion] [High automatic] [SMEs]
Tax incentives for R&D expenses (l. 140/97)	It provided tax incentives for R&D-related expenses. It had the ultimate goal of encouraging SMEs to make explicit their innovation activities	Tax exemption [Fiscal # low coercion] [High automatic] [SMEs]

Table 5.7 Business-oriented II—instrument mix

Instrument	*Content*	*Shapes Family Delivery Target*
Incentives for telematic connections in the textile, clothing, footwear industry (art. 103 (5))	Tax reduction aimed at incentivizing the development of telematic links, the speeding up of logistic flows, the exchange and acquisition of information, the creation of platforms for the development of standardized systems for the monitoring of the various phases of production and marketing It was reserved for the textile, clothing, and footwear sector in the attempt to increase the competitiveness of key areas of the "Made in Italy" initiative	Tax reduction [Fiscal # medium coercion] [High automatic] [SMEs]

(*continued*)

Table 5.7 (continued)

Instrument	*Content*	*Shapes Family Delivery Target*
Incentives for e-commerce (art.103, C. 5-6)	Tax credit for investment programs aimed at the development of IT solutions to carry out electronic commerce activities and digital training for staff. The eligible costs were related with hardware acquisition, software licenses, consultancy services, and advertising space online. The amount of the credit varied according to the beneficiaries. The minimum amount of the investment programs should be equal to €30,000, and the tax credit varied between 35 and 45% of total costs	Tax reduction [Fiscal # medium coercion] [High automatic] [SMEs]
Incubators for start-up (art. 106)	It provided technical assistance for the start-up of new innovative enterprises through qualified selected entities. The measure covered expenses related to the following activities: feasibility studies, infrastructures, organizational and financial assistance, training, and technical assessment of the projects. The measure aimed at encouraging the creation of business incubators that could favor the birth of innovative companies through the provision of services supporting the start-up phase	Consulting service [Information # medium coercive] [Mixed automatic] [Start-up]
State participation in risk capitals (art.106)	It provided state equity finance participation in risk capital for a minimum of 20% of the total capital of the enterprise. Alternatively, it could also provide funding to intermediary actors for the acquisition of a third of enterprise capital quotas (in this case no more than 50% of the total amount of the enterprise's capital). It was managed by the MISE. In order to apply for this aid, enterprises had to present a 3–5-year multiannual plan	Equity participation [Expenditure # high coercive] [Mixed automatic] [Enterprises+ HEIs+ PROs]
Investments in less-favored areas (art.8)	The incentive consisted of a tax credit that enabled firms to offset both direct and indirect taxes for investments on depressed areas. The incentive was exclusively devoted to companies carrying out specific types of activities (mining and manufacturing, service, tourism, trade and building activities, production and supply of electronic power, steam, and hot water, fishing and water farming, transformation of fishing and water farming products)	Tax reduction [Fiscal # medium coercion] [High automatic] [Enterprises]

Table 5.8 Business-oriented III—instrument mix

Instrument	*Content*	*Shapes* *Family* *Delivery* *Target*
Tremonti-Bis L. 383/2001 (art. 4-5)	It was a tax reduction for reinvested profits equal to 50% of the whole volume of investments in new instrumental goods exceeding the average amount of investments achieved in the last five years of taxation. It covered also expenses incurred in the training of personnel, which should be increased by no more than 20% of the cost of the remuneration due to the internal personnel during the training period	Tax reduction [Fiscal # medium coercion] [High automatic] [Enterprises]
Tecno-Tremonti (l. 326/2003 art.1)	It was a fiscal incentive for companies investing in R&D (product, process, and organizational innovation). The incentive was equal to 10% of the R&D costs from the taxable income. In addition, a further 30% deduction was applied to the difference between the R&D cost of the current fiscal year and the average R&D costs during the previous three fiscal years. Among the expenses were also included costs of ICT-related innovations. The same rule applied to SMEs that, pooling resources and creating consortia of at least ten enterprises, undertook innovative investments in information technologies	Tax reduction [Fiscal # medium coercion] [High automatic] [SMEs]
Creation of the Italian Institute for Technology (l. 326/2003 art.4)	Establishment of the Italian Institute of Technology (IIT). It was a foundation with the aim of promoting technological research and development of the country (mainly robotics, neuroscience, neuro technologies, computer science); jointly managed by the Minister of Economy and Finance and Minister for Higher Education and Research	Research institute [Regulation # low coercion] [Low automatic] [HEIs + PROs]
Credit securitization (l. 326/2003 art.2); D.M. MEF 16/09/2004)	Provided by the l. 326/04 and made operational in 2004. It reorganized the financing structure of R&I for the MIUR and the MISE. It foresaw the securitization of outstanding national credits in short-term financing by the two ministries	Administrative simplification [Regulation # medium coercion] [Low automatic] [HEIs + PROs + Enterprises]

(*continued*)

Table 5.8 (continued)

Instrument	*Content*	*Shapes Family Delivery Target*
Facilitated financing for ICT (FRI) (l. 311/04 art.4 C. 354)	This instrument provided subsidized loans in the form of capital advances by CDP (Cassa Depositi e Prestiti), a national investment bank, according to a multi-year repayment plan. The identification of the objectives and the different applications of the fund was defined in the National Research Plan and devoted to technological innovation, industrial sectors, tourism and craftmanship sector, plus agriculture and services	Cash advance Expenditure # high coercion] [Low automatic] [SMEs]
New measures to support the Italian economic system (l. 80/05)	Modification of the incentive system for enterprises in underdeveloped areas (l.415/92; 488/92). It basically substituted non-refundable grants with subsidized loans involving the participation of the banking system	Loan [Expenditure # high coercion] [Mixed automatic] [SMEs]

Table 5.9 Business-oriented IV—instrument mix

Instrument	*Content*	*Shapes Family Delivery Target*
National agency for the diffusion of technologies for innovation (l.266/05 art.1 C. 368d)	It carried out tasks related with the economic, financial, and scientific evaluation of industrial innovation projects. It promoted and coordinated activities aimed at forecasting technological-scientific and economic development trends. It operated on the basis of a three-year activity program updated annually, which determined its objectives and priorities	Advisory committee [Regulation # low coercion] [Low automatic] [HEIs + PROs + SMEs]
Tax relief for patent registration (l.266/05 art.1 C.351-352)	Tax exemption on patent fees	Tax exemption [Fiscal # low coercion] [High automatic] [HEIs + PROs + SMEs]

Table 5.10 Business-oriented V—instrument mix

Instrument	*Content*	*Shapes Family* *Delivery* *Target*
Tax exemption on capital gains from start-up (l. 133/2008 art.3)	It established that capital gains were exempted from personal income taxes when: they have been possessed by at least three years, they came from young companies (no older than seven years), and they were re-invested in the next two years in young start-ups that operated in the same sector as the first company. The maximum amount that could be tax-exempted was directly linked to a multiplier of the productive investments (material and immaterial goods and R&D expenses) carried out by the firm that originated the capital gain	Tax exemption [Fiscal # low coercion] [High automatic] [Start-up]
Italian investment fund	It was a private equity fund, dedicated to qualified investors, mainly financed by CDP and by other private banks. It was devoted to SMEs capitalization and grouping for specific development projects	Equity participation [Expenditure # high coercive] [Mixed automatic] [SMEs]
JEREMIE fund	It supported the access to finance for SMEs via structural funds interventions, by offering credit and risk funding to SMEs and a public guarantee to banks, in order to improve access to investment capitals by SMEs. It was devoted to activities aimed at developing new products, securing, and expanding market access	Equity participation [Expenditure # high coercive] [Low automatic] [SMEs]
National innovation fund	It acted as an instrument to reduce investment risk for banks and financial intermediaries that participated in the funding/financing of innovative projects based on the valorization and the use of patents. These resources aimed to help SMEs gain access to risk capital to finance innovative projects that made use of patents owned by SMEs	Economic guarantee [Expenditure #low coercive] [Mixed automatic] [SMEs]

(*continued*)

Table 5.10 (continued)

Instrument	*Content*	*Shapes Family* *Delivery* *Target*
Tax credit for firms financing research projects in universities and PROs (l.106/2011 art.1)	Aimed at promoting public research outsourced to the business sector. The allowance was equal to 90% of the incremental expenses incurred by the firm financing the outsourcing of research activities to PROs and universities	Tax reduction [Fiscal # medium coercion] [High automatic] [Enterprises + HEIs + PROs]
Venture capital stimulation	It aimed at stimulating venture capital creation and adoption to finance the development of start-ups, by providing that revenues for the capital invested were not subject to taxation	Tax exemption [Taxation #l ow coercive] [High automatic] [Start-up]

Table 5.11 Business-oriented VI—instrument mix

Instrument	*Content*	*Shapes Family* *Delivery* *Target*
Tax credit—skilled workers (L. 134/2012 art.24)	Tax reduction equal to 35% of the costs for employing highly skilled workers (individuals who had a PhD or a master's degree and were employed in R&D activities). The maximum amount of reduction for each enterprise was equal to €200,000 per year. Later, also extended to start-ups	Tax reduction [Fiscal # medium coercion] [High automatic] [Enterprises]
Fund for sustainable growth (L. 134/2012 art.23)	It was focused on technological innovation, and it substituted the former fund to sustain enterprises' investments (FRI) approved with the budgetary law for 2005. Managed by the MISE linked to Horizon 2020 guidelines and definitions. It simplified regulation and redefines the scope, the beneficiaries, and the mix of incentives available for indirect financing	Administrative simplification [Regulation #l medium coercive] [Low automatic] [Enterprises]

(*continued*)

Table 5.11 (continued)

Instrument	*Content*	*Shapes Family* *Delivery* *Target*
Creation agency for digital Italy (L. 134/2012 art.23)	Creation of a new agency for the implementation of the strategy for the digitalization of the country, managing funds for R&D projects based on ICT development. Suppression of the Agency for the Diffusion of Technology for Innovation, created with the budget law for 2006	Advisory committee [Regulation # low coercion] [Low automatic] [HEIs + PROs + enterprises]
Start-up law—legal framework for start-ups (l.221/2012 art.23)	For the first time an extensive regulatory framework has been arranged in favor of this type of company by legally defining their status according to some specific parameters: newly incorporated entities (< five years old), not distributing profits, dealing with the production, development and commercialization of innovative goods of high technological value. Its innovativeness was defined by the following criteria: at least 15% of company's expenses should be attributed to R&D, at least 1/3 of the total work force were PhDs, or alternatively 2/3 of the workforce must hold a master's degree, the enterprise was the holder depository of a license/ registered patent/software	Administrative simplification [Regulation # medium coercion] [Low automatic] [SMEs]
Fixed fiscal costs (l.221/2012 art.26)	Those companies which fulfilled the characteristics of a start-up were exempted from paying administrative fees related with the official registration of the company	Tax exemption [Fiscal # low coercion] [High automatic] [Start-up]
Tax exemptions for investments in start-ups (l.221/2012 art.29)	If taxpayers invested a share of their capitals, for a maximum of €500,000 for at least two years, in one or more innovative start-ups (directly or through collective investment saving organizations) they could benefit for four years of a deduction equal to 19% on personal income tax	Tax exemption [Fiscal # low coercion] [High automatic] [Start-up]

Bibliography

Banfi, A., Viesti, G. 2016. "Il Finanziamento Delle Università." In Viesti, G. *Università* in Declino. Un'indagine Sugli Atenei Da Nord a Sud, Donzelli, 277–304.

Bersani, Pier Luigi, Alfonso Lelio, and Letta Enrico. 2004. *Viaggio Nell'economia Italiana*. Donzelli.

Bologna, Federica, Angelo Di Iorio, Silvio Peroni, and Francesco Poggi. 2021. *Do Open Citations Inform the Qualitative Peer-Review Evaluation in Research Assessments? An Analysis of the Italian National Scientific Qualification*.

Borrás, Susana, and Charles Edquist. 2019. *Holistic Innovation Policy*. Oxford University Press.

Capano, Giliberto. 2008.55 Higher Education *Looking for Serendipity: The Problematical Reform of Government within Italy's Universities*.

Capano, Giliberto., and Elisabetta Gualmini. 2011. *Le Pubbliche Amministrazioni in Italia*. Il mulino.

Capano, Giliberto, Marino Regini, and Matteo Turri. 2016. Changing Governance in Universities: Italian Higher Education in Comparative Perspective.

Cobis, Fabrizio. 2008. I Nuovi Processi Di Innovazione e l'azione Del Ministero Dell'Università e Della Ricerca. *I nuovi processi di innovazione e l'azione del Ministero dell'Università e della Ricerca*: 1000–1005.

Coccia, Mario, and Secondo Rolfo. 2010. New Entrepreneurial Behaviour of Public Research Organisations: Opportunities and Threats of Technological Services Supply. *Int. J. Services Technology and Management* 13 (1/2): 134–151.

Corradi, Fiammetta. 2009. *Alla Ricerca Dell'eccellenza: Le Politiche per l'eccellenza Nell'istruzione Superiore in Quattro Paesi Europei*. LED, Edizioni universitarie di lettere, economia, diritto.

De Maio, Adriano. 2011. *L' Innovazione Vincente*. Brioschi.

Di Maio, Michele. 2014. The Italian Economy, the Economic Crisis and Industrial Policy. In *Structural Change, Competitiveness and Industrial Policy: Painful Lessons from the European Periphery*, ed. A. Teixeira and E. Silva. Routledge.

Donina, Davide, Michele Meoli, and Stefano Paleari. 2015. Higher Education Reform in Italy: Tightening Regulation Instead of Steering at a Distance. *Higher Education Policy* 28 (2): 215–234.

European Commission. 1999. European Trend Chart on Innovation Country Report: Italy.

———. 2000. European Trend Chart on Innovation Country Report: Italy.

———. 2001. European Trend Chart on Innovation Country Report: Italy.

———. 2002. European Trend Chart on Innovation Country Report: Italy.

———. 2003. European Trend Chart on Innovation Country Report: Italy.

———. 2004. European Trend Chart on Innovation Country Report: Italy.

———. 2005. European Trend Chart on Innovation Country Report: Italy.

———. 2006. European Trend Chart on Innovation Country Report: Italy.

Gallo, Riccardo. 2016. *Torniamo a Industriarci: A Novant'anni Dalla "Grande Crisi"* Guida.

Gallo, Riccardo, and Francesco Silva. 2006. *Le Condizioni per Crescere: Diagnosi e Proposte per Il Sistema Produttivo*. Il Sole 24 ore.

Gallo, Riccardo, Mauro Mallone, and Vincenzo Zezza. 2008. Technology Transfer: The RIDITT Experience. *L'industria* 1: 207–226.

Lai, Luigi. 2016. Innovation Policies and SMEs Development. In *Entrepreneurship: Antecedents and Effects ("Przedsiębiorczość Międzynarodowa")*, ed. M. Kosała, M. Urbaniec, and A. Żur, vol. 2, 2nd ed., 147–163. Kraków: Cracow University of Economics.

Lanza, Orazio, and Kostas Lavdas. 2000. The Disentanglement of Interest Politics: Business Associability, the Parties and Policy in Italy and Greece. *European Journal of Political Research* 37 (2): 203–235. https://doi.org/10.1023/A:1007017624486.

Lissoni, Francesco, and Fabio Montobbio. 2015. The Ownership of Academic Patents and Their Impact. Evidence from Five European Countries. *Revue Economique* 66 (1): 143–171.

Lissoni, Francesco, Peter Lotz, Jens Schovsbo, and Adele Treccani. 2009. Academic Patenting and the Professor's Privilege: Evidence on Denmark from the KEINS Database. *Science and Public Policy* 36 (8): 595–607.

Lucchese, Matteo, Leopoldo Nascia, Mario Pianta, and Mario Pianta. 2016. Industrial Policy and Technology in Italy. *Economia e Politica Industriale* 43 (3): 233–260.

Malerba, Franco. 1993. The National System of Innovation: Italy. In *National Innovation Systems: A Comparative Analysis.*

Ministero dell'Istruzione dell'Università e della Ricerca. 2005. *Programma Nazionale per La Ricerca.*

MISE. 2010. *Gli Incentivi Alle Attività Produttive. Effetti Della Legge 46/82 Sull'innovazione e Sull'industrializzazione Dei Prototipi Di Ricerca.*

Nascia, Leopoldo, and Mario Pianta. 2013. *ERAWATCH Country Reports 2013: Italy.*

Nascia, Leopoldo, Bianca Potì, and Emanuela Reale. 2012. *ERAWATCH Country Report 2012: Italy.*

Pierantozzi, Valentina. 2007. *ERAWATCH Analytical Country Report 2007: Italy.*

Potì, Bianca, and Emanuela Reale. 2006. *Project Funding. Italy.*

———. 2007. Changing Allocation Models for Public Research Funding: An Empirical Exploration Based on Project Funding Data. *Science and Public Policy* 34 (6): 417–430.

———. 2010. *Erawatch Country Reports 2010: Italy.*

———. 2011. *ERAWATCH Country Report 2011: Italy.*

Poti, Bianca, Emanuela Reale, and Valentina Pierantozzi. 2008. *ERAWATCH Country Report. An Assessment of Research System and Policies. Italy.*

Reale, Emanuela, and Lucio Morettini. 2017. *Public Funding Country Profile Italy Annex 11.*

Reale, Emanuela, and Bianca Potì. 2009. Italy: Local Policy Legacy and Moving to an 'In Between' Configuration. In *University Governance: Western European Comparative Perspectives*, ed. Catherine Paradeise, Emanuela Reale, Ivar Bleiklie, and Ewan Ferlie, 77–102. Springer Verlag.

Silva, Francesco. 2007. La Politica Industriale Del Governo Prodi. *ECONOMIA E POLITICA INDUSTRIALE* 2: 7–19.

Sirilli, Giorgio. 2010. *La Produzione e La Diffusione Della Conoscenza. Ricerca, Innovazione e Risorse Umane.*

Spence, R.E. 2014. "Italy." In *Chandler, J. A. (2014). Comparative Public Administration (Second Editon.). Routledge, Taylor & Francis Group.*

Traù, Fabrizio. 2009. Logica Di Un Sistema Di Sostegno Alle Imprese. La Politica Industriale Italiana Dal Secondo Dopoguerra a 'Industria 2015'. *ECONOMIA E POLITICA INDUSTRIALE* 4.

CHAPTER 6

Comparative Research and Innovation Policy Design

This study explores how governments design national R&I policy strategies and the influence of different governance configurations on policy design choices. This means investigating policymaking both internally—focusing on the institutional coordination of sectorial responsibilities—and externally, looking at the relationship between policymakers and policy takers. The book explores the role of different partisan political orientations in shaping policymakers' choices for alternative instrument features. Then, narrowing the focus on those actors involved in the policymaking of national R&I strategies, it concentrates on the different ways institutional structures, and their related coordination challenges, influence policymaking activities. Finally, it zooms in on the interactions between policymakers and policy takers, investigating how their configurations influence policy design choices. Inspired by Lasswell's definition of politics as "*the process of who gets what, when and how,*" this research investigates the influence of politics, institutional arrangements, and interest intermediation strategies on R&I policy design.

The innovative contribution of our research is twofold. First, it analyzes the determinants of policy design choices from the perspective of both the actors "making" and the actors "receiving" these decisions. This represents a novel contribution to the literature because it combines the supply-side with the demand-side of policy design. This perspective is especially relevant for R&I policies, where policy design choices encompass

C. Acciai, *Policy Design for Research and Innovation*, International Series on Public Policy,
https://doi.org/10.1007/978-3-031-36628-4_6

interdependent issues related to knowledge production and exploitation practices, which do not necessarily map easily onto the typical remits of formal policymaking institutions (Pelkonen et al. 2008). Secondly, to interpret this biunivocal relationship, the author theorizes a new R&I policy instrument treatment relying on instrument families, shapes, and delivery components. This classification differentiates instruments according to the behavioral changes required for policy takers and it illustrates the different ways in which public authority is exercised. It allows for treating policy instruments as the operational dimensions of different governance arrangements (Capano et al. 2019) and investigating the pathways shaping the politics of instrument choices and their blending into different mixes.

Focusing on actors, their interactions, and different stakes, the book unravels the (sectorial) complexities characterizing R&I decision-making while exploring, from an actor perspective, how policy instruments are selected and blended. This will contribute to a deeper understanding of how governments design more ambitious policy mixes aimed at solving the complex, and cross-sectorial in nature, social challenges contemporary societies are facing. The cross-sectorial nature of R&I policy challenges implies that, in addition to the internal ability of policymakers (e.g., coordination of ministerial responsibilities), we must also consider their capacity to account for the characteristics of policy takers, which will bring with them specific sets of associated interests, problem perception, and configurations related to the tradition of the sectorial policies involved. Policy recipients also matter, but due to their heterogeneity in the R&I sector we still know little about their strategic behaviors. For all these reasons, it becomes relevant to investigate the way target groups relate to the policy process and the actors involved in these dynamics. This perspective helps understand how actors interact for the selection of policy instruments, and the way this process influences the diachronic evolution of policy mixes. Moreover, the comparative analysis between Italy and France sheds light on national policy design practices, showing the strategies and legacies shaping R&I instrument blends, by providing a comprehensive historical overview of the instruments adopted. Since instruments represent the operational dimension of governance arrangements (Capano et al. 2019), by following their evolution, we shed light on both the relationship between policymakers and target populations and the internal dynamics shaping the evolution of national R&I governance arrangements. We aimed first at exploring the variations in R&I policy mix (by testing our new instrument classification typology) then, through an analysis of the way actors interact, we identified the pathways which can help to make sense of the political process behind the selection of alternative instrument mixes. Indeed, the novelty of our contribution rests precisely on explaining how tools are selected from the perspective of the actors making the decisions. These two countries share many similarities in their aggregate policy mix characteristics. However, when we look at the aggregate characteristics of how different instruments exercise social control (instrument shapes) and the relationships between policymakers and target populations (delivery structure), the results display a greater variety. Such differences reflect alternative approaches the two countries have undertaken to interact with target populations, as well as the different degrees of political entrepreneurship and organizational capacities of national R&I performers.

This conclusive chapter is organized as follows. In the first section, we will briefly summarize the most relevant historical events characterizing the evolution of the national R&I sector in our two cases, as extensively discussed in Chaps. 4 and 5. In Sect. 6.2 we will discuss the politics of R&I policy design, presenting evidence from the two case studies. The chapter continues with the analysis of the institutional intermediation and interest configuration strategies. It concludes with a final discussion and a final presentation of the policy implications and limitations of the study.

6.1 A Binary Evolution in R&I Policy Design: The Case of France and Italy

Despite some differences in the political choices that characterized their national strategies, French and Italian R&I policy design practices represent two highly comparable cases. Both countries have a long history of state-led enterprises in highly intensive R&I sectors, accompanied by public research organizations playing a pivotal role. At the beginning of the 1990s, both cases went through a process of disposal of state participation in many sectors of the national economy. In Italy, due to the absence of clear sectorial orientations, the system endogenously tended to favor those actors who were able to autonomously lead and design innovation projects—especially those actors who were able to bear the costs related to delays in the instrument activation process and funding provisions. This is basically why Italy is said to display severe weakness in its capacity to adopt mission-oriented policies (Onida and Malerba 1990), vulnerabilities that are also well recognized in official strategic documents (Grilli and Mariotti 2006). On the other hand, in France, the dismissal process took a different path, and policymakers succeeded in maintaining a greater degree of participation in specific sectors of the national economy. As argued by one of our interviewees, in the case of the nuclear industry, the State didn't go through a process of economic disengagement, because it wanted to preserve strategic autonomy in this sector. There is a widespread agreement about the political willingness to preserve state competences in specific industries. This is strategic from a geopolitical and international perspective, although this is not the "classic" type of state control, but it is rather a supervision role (INTERVIEWEE 17).

Figure 6.1 chronologically displays the main events characterizing the history of the R&I policy sectors' evolution in our two cases.

In 1998, both countries engaged in a systematic examination of their national R&I systems. This happened with the report to the parliament of the minister of higher education and research in the Italian case, and with the *Guillaume* report in France. In both instances, these documents triggered a reorganization of national (and regional) R&I institutions, together with a restructuring of their national economic development strategies. In France this process culminated in the drafting of the 2003

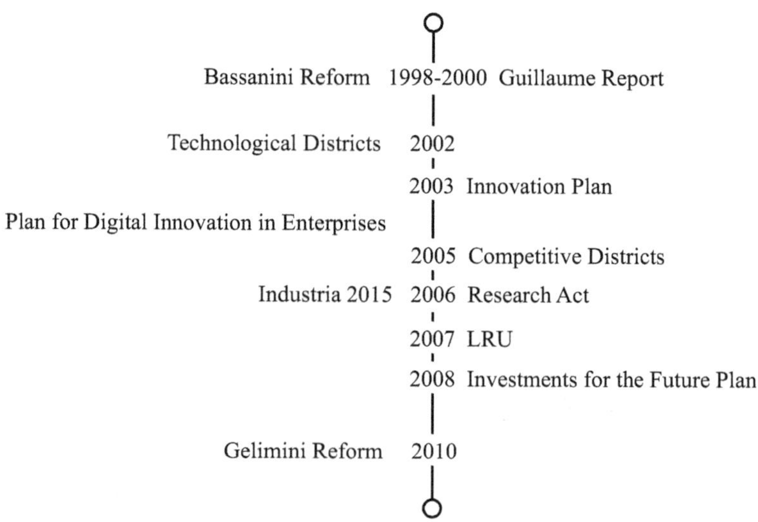

Fig. 6.1 The chronology of the two cases

Innovation plan, followed by the Research Act in 2006 and the Law on the Freedom and Autonomy of the University, in 2007.[1] Differently, for the Italian case, the revision of national R&I governance found its place in the broader framework of a country-wide administrative reorganization (the *Bassanini* package). Although this reform defined some of the underlining principles influencing the evolution of national R&I, it didn't succeed in producing a system of structured reforms, as happened in the French case. However, looking at the chronology of national interventions across the two cases, it is still possible to draw some parallels between the evolution of the French and Italian R&I systems.

In both "*Industria 2015*" and the "Investment for the Future Plan," national governments, either through the collaboration between ministers or with the creation of a new organization (the French General Commissariat for Investment), identified specific technological and productive areas for R&I investments while designing long-term strategies for national development. These are important instances of national government's willingness to maintain the responsibility of translating national objectives into R&I objectives. However, the Italian experience had a short life because this strategy was modified by a newly elected cabinet (2008), which reverted the (soft) mission-oriented nature of the policy into a diffusion-oriented logic (Traù 2009). While in the case of the Investment for the Future plan the strategy was slightly reoriented and re-calibrated, it nevertheless did manage to maintain a mission-oriented aim across different political mandates. Both countries experienced similar interventions in the field of competitive clusters. In this sector Italy was the forerunner; since the beginning of the 1980s the organization of the traditional sectors of production in the northern regions had progressively evolved toward a system based on districts. Over the years, this format was expanded and consolidated—also in other areas of the country. This was accompanied by specific funds (e.g., fund for underutilized areas and European structural funds) and legal provisions to simplify the division of competences between regional and national authorities. The same policy design was also replicated in 2002 by the Ministry of Higher Education and Research, which established the "technological districts." This system shared many similarities with industrial districts. However, contrary to the districts which were mainly based on SMEs, it expanded the pool of potential beneficiaries and identified a clearer mission of promoting highly

[1] We already discussed in Chap. 3 the timing and the drafting of these reforms.

qualified human capital coming from public research centers. The French cluster experience started in 2004 with the "*pôles de compétitivité*," territorially based competitive clusters fostering the synergetic development of specific sectors (or technologies). These networks were territorially defined by the coexistence of a highly specialized industrial basis matched with research and education organizations that, by means of partnerships, could have access to funding under privileged conditions and discounts on corporate and social taxes (European Commission 2005). In 2008, the *pôles de compétitivité* became one of the main recipients of a public procurement-based instrument mix. Contrary to the Italian case, the French industrial clusters were almost directly steered by the central government, and the local dimension was only the most suitable administrative level to implement such a strategy. Therefore, once again, Italy and France went through very similar policy design experiences. In the first case, a national framework was missing, and strategic choices were mainly left in the hands of regions, while in the second, clusters were mainly governed by the central state. Finally, both countries implemented a reform of national higher education systems. The French reform, the Law on the Autonomy and Responsibility of Universities (LRU), reduced the size of universities' administration councils, while increasing the role of their presidents. Universities acquired greater budgetary and financial management competences of their institutions, and they were required to sign a four-year contract (*contrats quadriennaux*) with the Ministry of Higher Education and Research for coordinated management of their institutional strategies. The logic of the LRU was to put universities at the center of the system as the main operators of public research. This ran in parallel with the gradual decrease in power for public research organizations (PROs) regarding both funding capacity—with the creation of a National Research Agency (ANR)—and an evaluation role, after the establishment of the National Evaluation Agency (AERES) (Musselin 2017). To some extent the LRU was the last piece of a bigger design to hollow out PROs' historically consolidated powers in the national governance of R&I. As we discuss in the following paragraphs, the extent to which these interventions have been successful in overcoming institutional friction is debated. Nevertheless, as demonstrated by a strong mobilization of researchers, this has been a reason for political conflict between R&I policymakers and policy takers. In the same vein, the 2010 higher education reform in Italy (*Riforma Gelmini*) arose in a general climate where universities were perceived as "enjoying autonomy, without responsibility," in which pressures of similar reforms previously applied in other European countries were adopted to promote similar changes (Capano 2008). In line with other higher education reforms in Europe, policymakers aimed at transforming Italian universities into organizations inspired by a managerial-oriented logic, with a specific focus on research and teaching performance, for securing their public funding allocations. The governance of this sector has been historically characterized by a dynamic of push and pull between the State and universities. Policymakers have tried to implement different rationalization strategies in the sector, while universities have tried to resist in maintaining their traditional model refuting the push toward increasing autonomy embedded in many reforms (Capano et al. 2016; Reale and Potì 2009). Similarly to France, the law modified the internal governance structure of Italian universities, the provisions regarding recruitment, as well as the power and competences of internal university organization structures. The two reforms shared many similarities (e.g., increasing financial and managerial autonomy of universities; and promoting a push toward a competitive-based funding provision), and they both represented an example of external public regulation in governing the national higher education sector.

6.2 The Politics of Instrument Choices: A Comparative Perspective

Policy design choices reflect political and social values of policymakers, as well as national trends and ideas about "good" policies (Schneider and Sidney 2009). The framings of policy problems, and their solutions, tend to cluster over time into long-term policy rationalities (Edler et al. 2016; Fagerberg 2017), which ultimately act as a lens filtering information and directing policymakers' attention toward different issues (Wilson 2000). We know that different political contexts can influence the availability of financial resources for investments in R&D in various ways (Bergek et al. 2015). And that the deployment of R&I instruments is able to alter the space of politics (Edquist and Borrás 2013). However, we still know little about how different political priorities are translated into R&I instrument mixes; hence, whether specific instrument blends are preferred by governments of different political orientations. Policy instruments are chosen with a purpose (Edler et al. 2016); the way these are designed and implemented represents a manifestation of different political views. Therefore, we expected that specific R&I instrument blends would be associated with different political coalitions.

As discussed in previous chapters of this book, instruments can be grouped into families according to the degree to which governments use their power, yet each of them displays a high level of variation in the type of relationships they establish with recipients. For, in addition to social control, any given tool embeds a particular way of exercising it, influencing how target populations behave by privileging some actors and interests over others (Kassim and Le Galès 2010; Lascoumes and Le Gales 2007). The proposed instrument typology (family-shape-delivery) allows a deeper understanding of how social control is exercised, providing a greater granularity to the study of the politics behind R&I policy design choices.[2] Figure 6.2 describes the distribution of R&I policy instrument choices by cabinet political orientation across the two cases. It focuses on instrument families, hence on the means-end rationality shaping the authoritative relationships between decision-makers and recipients.

Italy shows some homogeneity between left- and right-wing oriented cabinets. In both cases expenditures and fiscal instruments represent the highest share of the total mix, with smaller variations in the use of regulation and information instruments. The Monti government, whose peculiar nature is discussed in Chap. 5, only focused on fiscal and regulation instrument types. France shows a clear preference for expenditure instrument types, which cut across opposite political orientations. However, the

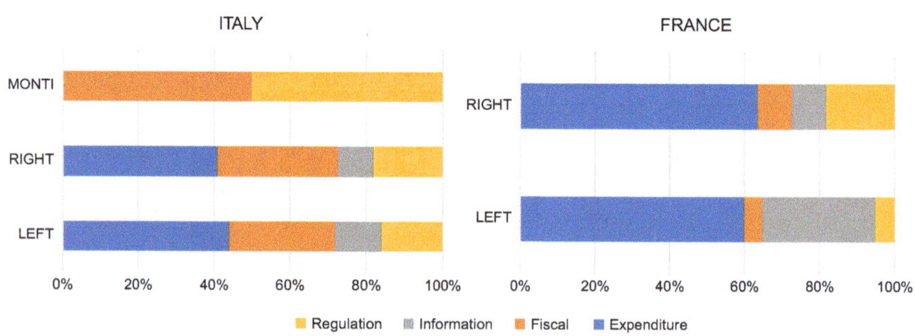

Fig. 6.2 Distribution of instrument families by cabinet political orientation

[2] The instruments used for the following analysis are all the instruments implemented in the two case studies, as described by the tables in Appendix of Chaps. 4 and 5.

remaining share of instrument choices displays higher polarization between different political orientations. Left-wing governments show a clear tendency toward information instruments, which is replaced by a mix of regulation and fiscal instruments for the right wing. Both countries show a greater dominance of expenditure instruments, which is in line with the investment-based nature characterizing R&I policy. However, only by adopting the analytical differentiation between fiscal and expenditure instrument families, as proposed by our classification, we can highlight how the two countries undertook different strategies in the way public authority is exercised. Indeed, the greater share of fiscal instruments adopted in Italy highlights a different political economy in tool choices, which would otherwise be very similar, especially for the two right-wing-oriented national cabinets.

Figure 6.3 explores the distribution of coerciveness across the different instrument shapes selected. This variation describes the characteristics of the substantive behavioral requirements administered to the target population to obtain compliance or to deliver expected outcomes.

In this case Italy shows a greater variety of instruments across the national political spectrum. Right-wing cabinets display larger preferences for highly coercive instruments at the expense of less coercive ones. While, on the other hand, their political opponents display a completely inverted set of choices. France displays a greater similarity across the political spectrum in the set of selected R&I instruments. In comparison with the Italian case, it is remarkable to note the high share of more coercive instruments across the political continuum.

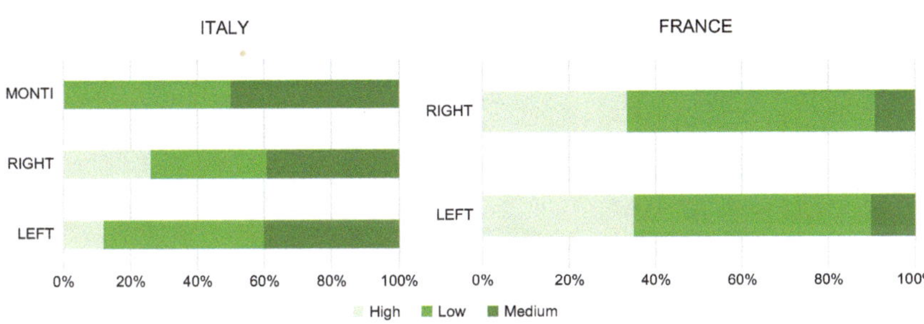

Fig. 6.3 Distribution of instrument shape coerciveness by cabinet political orientation

Figure 6.4 describes the last component of the policy instrument classification presented in this study: the instrument delivery. This component describes how, in addition to social control, every tool embeds a particular way of exercising it by privileging some actors and interests over others (Kassim and Le Galès 2010). It is classified according to different degrees of automaticity, attributing a fine-grained contribution to the role of actors engaged in procedures following policy instrument formulation, and their power to steer their enactment.

The Italian case shows a high degree of differentiation across the political spectrum. Left-wing coalitions show a clear preference for low automatic delivery, which relies on administrative and/or organizational structures. Conversely, right-wing parties tend to prefer higher automatic delivery, which makes use of already-existing structures for their implementation, although the difference is minimal. This is the first instance of greater political heterogeneity in France. Despite the high share of low automatic instrument choices across the continuum, which resonates well with the high authoritative attitudes displayed in Fig. 6.2, the political divide is clear in the case of mixed delivery instruments. So, the instrument components relying on the cooperation with R&I stakeholders are preferred by left-wing governments. The two countries show a different distribution of delivery choices both across the political spectrum and at the aggregate level.

In a comparative perspective, Italy tends to be slightly more polarized in the choices of R&I instrument blends, compared to France. However, both countries show, on average, a low degree of political polarization in instrument choices across the political spectrum. This suggests that our expectations about the association between R&I instruments' blends and

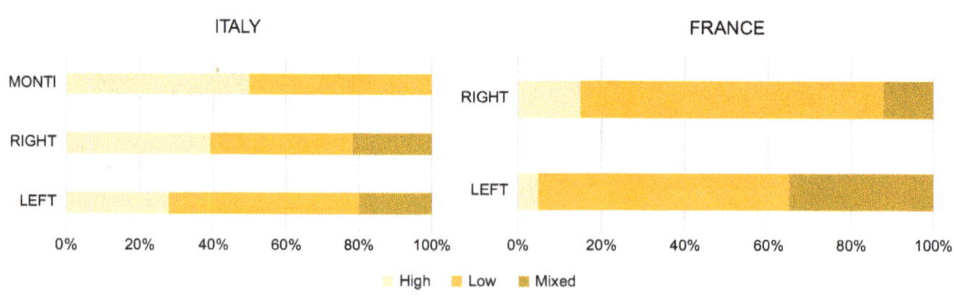

Fig. 6.4 Distribution of delivery component characteristics by cabinet political orientation

different political coalitions in office were not confirmed. Consequently, we can conclude that these choices are only an extremely moderate manifestation of different political orientations of the cabinet in office.

6.2.1 Path Dependency and Political Turnovers

The general distribution of instrument choices across the two cases displays some consistent patterns. As discussed in greater detail in Chap. 3, we expected to identify different trends in the distribution of government choices for R&I policy instrument features across countries characterized by different R&I policy-type traditions. The literature tells us that problem-solution framings tend to cluster into long-term paradigms, acting as a lens that filters information and focuses attention on specific issues (Wilson 2000). These mechanisms should therefore lay the foundations for developing ways of designing policies that consolidate into specific policy types.

France and Italy have followed a binary path of evolution in their R&I policy choices. However, if we look at the specific dynamics characterizing these developments, their constitutive differences seem more evident. The former shows clearer preferences for expenditure instrument families, higher coercive shapes, and low delivery structures that rely on existing bureaucratic structures for the implementation of public action. These features are in line with the national tradition of R&I mission-oriented types of policies. Indeed, France is historically characterized by a history of large-scale industrial programs, where governments used to have primacy in the selection and support of specific technological sectors. Here, policies are distinguished by a high degree of centralization in the determination of goals (Foray 2009; Foray et al. 2012). The State is closely involved in financing and executing R&I, and this characterizes a fairly centralized system which is able to ensure stability in technological choices (OECD 2014). Conversely, Italy shows a greater share of preferences for fiscal instruments, lower coercive shapes, and higher automatic delivery components. This pattern of choices is highly consistent with a diffusion-oriented R&I policy type whose instrumentation relies on less authoritative and more automatic instrument mixes. In this context, policy decisions tend to be taken in a decentralized manner, and the implementation of these tasks is delegated to either industrial associations or cooperative research organizations (Ergas 1987). This leaves policy takers with greater autonomy, but also responsibility, in the design of their innovation strategies (Evangelista 2007; Gallo and Silva 2006).

Therefore, France and Italy show different consolidated patterns of instrument choices. These reflect both national political priorities (how different political preferences have been translated into R&I issues) and different consolidated policy traditions characterizing the way authority is exercised and the relationship with policy takers.

Policy instruments embed different typologies of an authoritative relationship between decision-makers and policy recipients. These represent different means-ends rationalities of the inducement implemented by public action. All of this is reflected in the instrument mix implemented by different countries. For instance, in the French case, the high share of preferences for more authoritative instrument mixes is in line with the establishment of new agencies like AERES and ANR. In the Italian case, although the preferences are more scattered and less consistent across policy types, it is possible to identify some similarities with the diffusion-oriented type. Especially for what concerns the focus on fiscal instruments and automatic delivery, that is in line with the large support for SMEs. However, compared to the French case, Italy displays less harmony in the choices of R&I instruments, and this can be explained by different factors. Italian politics has never fundamentally considered these themes as central for the development of the country (INTERVIEWEES 11-7-10). Furthermore, *"nowadays Italy has a very efficient R&I instrumentation, we do really have all we need. What is missing, is exactly the idea, a political leadership eager to invest in this sector, which will also be able to drive the industrial sector to mobilize"* (INTERVIEWEE 10). National politicians are not clear about where they want to go and how they want to get there. Instead of starting from a political vision on where to go, policymakers tend to begin from instruments because they can be more convenient, or they are simply eager to fulfill patronage dynamics, but on the other hand they do little if there is no idea behind them (INTERVIEWEE 8). Political ideology doesn't seem to matter much, and policy mix characteristics seem to be rather a matter of satisfying the priorities of different political constituencies (INTERVIEWEE 1). Such a narrative on electoral constituencies acquires greater relevance if we consider that some instruments have been activated only for a specifically short period of time (like the case of one-off implementation of fiscal instruments), as if they were only meant to please the will of a specific electorate in a specific window of time. Consequently, on certain issues related with economic, industrial, and knowledge development of the country, there is no continuity from a political perspective even though, theoretically, on these issues they should all agree (INTERVIEWEE 8). These trends, matched with the instability of Italian governments, especially in comparison with the French case, can be interpreted as one of the triggering causes of the failure of the Italian political system to provide sound governance, also in the case of R&I policies (Spence 2014).

6.3 Institutions: Ministerial Organizational Layouts

From an institutional organizational perspective, R&I policies like, for example, climate policies (Van Asselt et al. 2015), migration policies (Scholten et al. 2017), and health care policies (Trein 2017), stand at the intersection of "classic policy sectors." In these policymaking contexts the burden of design choices is increasingly shared across different policymakers. This requires the institutional and organizational capacity necessary to cope with issues at the intersection of "classic policy sectors," a practice that often comes together with an increasing blurring of competences between executive national departments. Consequently, to gather the necessary resources and to design holistic R&I interventions, new institutional configurations have to be found among different policy responsibilities (Borrás and Edquist 2019).

The way policymaking responsibilities are organized reflects national specificities about a country's institutional organization and the framings of its national R&I policy strategies. Since ministerial organization is not strictly determined by law, for the most part, these structures tell us how national governments framed policy problems differently and which political priorities must be tackled (Peters 1998). Our contribution builds on this evidence and investigates how different morphologies of ministerial organization practices influence the ability of policymakers to design an instrument mix capable of considering multiple traditional subsectors of public action.

France and Italy have both pursued an external R&I institutional specialization pattern. Policy competences lie under the remit of the minister for higher education and research and the minister with economic competences (respectively called the minister of economic development in Italy and the minister of economy and finance in France). Consequently, since the success of this institutional structure is highly dependent on the ability to create dialogue among different locus of policy responsibilities, we expected that countries undertaking similar ministerial specialization strategies would likely adopt similar coordination practices (informal institution).

Policymakers tend to perceive their responsibilities according to a linear understanding of the innovation process, where one minister (the one in charge of research and higher education policies) deals with research, a second (the one in charge of industrial and economic development policies) has the titularity over the experimental development phase and eventually its market applications (INTERVIEWEES 5-21-17). Accordingly, different ministers with integrative R&I competences are also perceived as having two different and separate constituencies. Where the minister with responsibilities on research policies refers to public research organizations and universities, the minister with competences on economic development interventions is mainly oriented toward helping enterprises and the innovations developed by them (INTERVIEWEE 17). Therefore, given the combination between different institutional perspectives on R&I issues and the different constituency-based nature of this relationship, coordination between policymakers has not always been a smooth process.[3] Table 6.1 summarizes the coordination practices implemented across the two cases in the attempt to overcome sectorial-based R&I decisional deadlocks.

In the French case there has been a first attempt of coordinating policymaking with the Civil Budget for Research and Technological Development (BCRD) where, despite the role of formal coordinator attributed to the minister for higher education and research, different functional ministers preserved enough power to develop their own strategy. This sometimes nullified the cooperative effort of the whole coordination platform (Cytermann 2006). Similarly, in the Interministerial Mission for Research and Higher Education (MIRES), despite the introduction of an integrated monitoring, each functional minister preserved its own budget and competences over the design of individual research strategies (OECD 2014). The resulting strategy demonstrates how the coordination between ministers remained largely unstructured, simply

[3] Here coordination is intended as the spectrum of activities in which one party alters its own political strategy to accommodate the activity of others in the pursuit of similar goals (Zafonte and Sabatier 1998, p. 480).

Table 6.1 Ministerial coordination initiatives: France and Italy

France	*Italy*
The Civil Budget for Research and Technological Development—BCRD—(1997)	**National Research Program—PNR—(1998)**
Budgetary platform autonomously led by the minister of higher education and research. It combined the credits for public research within a single procedure, also providing sectorial ministers with the possibility to participate in the budget. However, its relevance as an instrument for interministerial coordination should not be overestimated. Indeed, when a specific research sector was deemed as strategic for a given ministry, it had the possibility to negotiate directly with the minister of finance	Programmatic document defining national R&I strategies and activities to be promoted (produced almost every three years). It is formulated by the minister for higher education and research after extensive consultations with the actors of the innovation system
Constitutional By-law on Finance—LOLF—(2006)	**Interministerial Committee for Economic Planning—CIPE—(2006)**
It provides a common framework for budgeting procedures in R&I by creating an Interministerial Mission for Research and Higher Education (MIRES), widening the scope of the BCRD. In this framework, the minister responsible for research oversees the coordination of government action among different ministers. However, this didn't provide any obligation for joint programming, so each minister could participate in the mission simply by including its autonomously designed program	National committee in charge of coordinating and planning national economic policies. It represented an additional forum, a director table, where ministers with R&I competences could gather and discuss national strategies (especially funding allocations) under the coordinating role of the prime minister and the minister of economy and finance. In 2006 it has been transferred under the prime minister's office, to strengthen its role as coordinator, while re-centralizing the control over the interministerial policymaking practices

(*continued*)

Table 6.1 (continued)

France	*Italy*
National Strategy for Research and Innovation—SNRI—(2008) Bottom-up priority-setting process involving public research, business, and civil society stakeholders, to identify the main social challenges national R&I policies should be directed at. The Investment for the Future Plan (for the first wave of applications) was in line with these priorities **Commissariat Général à l'Investissement (2010)** National Committee with the task of managing the implementation of the Investment for the Future Plan. It coordinates interministerial policy design under the authority of the PM as well as the cooperation among other governmental bodies responsible for the distribution of funds. It ensures the transparency and quality of the selection procedures, the allocation of funds to existent operators, as well as the overall coherence of the strategy	

replicating the competences of the ministers involved (Cytermann 2006). As summarized in Table 6.2 below, which describes the MIRES for 2012, the minister of higher education and research was the major contributor to the mission, by funding 50% of the programs included in the strategy.

Program 172 funded PROs[4] through institutional block funding, the National Research Agency through competitive funding, as well as tax incentives schemes like the Tax Credit for Research. Program 150 financed only higher education institutions under the supervision of the minister of higher education and research. However, the higher education sector was also a highly contended area between different ministries. Indeed, some schools, for example, *Grandes Écoles*, fell under the supervision of other

[4] Like the French National Centre for Scientific Research (*Centre national de la recherche scientifique*, CNRS), the French National Institute of Health and Medical Research (*Institut national de la santé et de la recherche médicale*, INSERM) and the The National Institute for Research in Digital Science and Technology (*Institut national de recherche en sciences et technologies du numérique*, INRIA).

Table 6.2 The ministerial contribution to MIRES (2012)

Program	*Title*	*Funding ministry*
142	Higher education and agricultural research	Ministry of Agriculture, Food, and Forestry
186	Cultural research and scientific culture	Ministry of Culture and Communication
190	Research in the fields of sustainable energy, development, and planning	Ministry of Ecology, Sustainable Development, and Energy
191	Dual-use research	Ministry of Defense
192	Research and higher education in economic and industrial fields	Ministry of Economy and Finance
150	Higher education courses and university research	Ministry of Higher Education and Research
172	Multidisciplinary scientific and technological research	Ministry of Higher Education and Research
187	Research in environmental and resource management	Ministry of Higher Education and Research
231	Student life	Ministry of Higher Education and Research
193	Space research	Ministry of Higher Education and Research

functional ministers (e.g., the *Écoles des Mines* was under the supervision of the Ministry of Economy and Finance), who were extremely reserved against any attempt at interference by the minister of higher education and research (Cytermann 2006, p. 88). The functional ministers involved were mainly institutions detaining joint supervision of PROs (e.g., the minister of agriculture with the National Institute for Agricultural Research; the minister of economy and finance with the National Institute for Research in Digital Science and Technology). Disaggregated data by each program were not available. However, Table 6.2 clearly describes the nature of the ministerial-based programs and the thematic-based division of competences between ministers, demonstrating a persistent sectorial-based ministerial struggle for power.[5] The binding powers of the National Strategy

[5] When presenting the MIRES architecture to the Higher Council for Research and Technology (CSRT), an observer noted a certain correspondence between the programs of the other ministers and the major technical bodies of the state (*Corps d'état*): "energy and industrial research" *Corps de Mines*, "Equipment transport and habitats" *Corps des Ponts*, "Dual research" *Corps des ingénieurs de l'armement*, "Higher education and Agricultural Research" *Corps du Génie rural et des Eaux et Forêts* (Cytermann 2006, p. 87). A relevant detail for understanding the whole picture is the fact that many of the personnel in the *Corps d'état* are people coming from a *Grandes Écoles* training.

for Research and Innovation were also quite limited. The strategy was not set at the operational level, it didn't allocate a budget, and its action was mainly focused on the provision of thematic guidelines for R&I national themes (OECD 2014). Under the Sarkozy presidency, within the framework of the Investment for the Future plan, a new layer was added to the national governance of R&I, through the creation of the General Commissariat for Investments (CGI). This committee represented a clear attempt to re-centralize some of R&I policymaking competences under the authority of the prime minister.[6] Therefore, its contribution to the pursuit of coordination in R&I policy responsibilities was factually limited. The CGI managed the enactment of the Investment for the Future plan, but it did not implement any explicit attempt at coordination or interaction with the other ministers involved in the national R&I effort (Lepori et al. 2017).

Differently, Italy did not implement any common budgeting procedure as a strategy to pursue coordination among ministers with R&I competences. The National Research Plan (NRP) was limited to the identification of broad national objectives, without providing any specification regarding the effective implementation of different measures. On top of that, the lack of financial commitment to the multiannual strategies defined by the document has often resulted in unrealistic targets (Nascia and Pianta 2013, p. 24) and unsatisfactory implementation practices (Grilli and Mariotti 2006). The formal titularity of the minister for higher education and research in drafting this programmatic document did not provide this actor with any coordination power. The centralization of competences was mainly formal, and it did not provide any effective result in terms of coordination among the actors involved (INTERVIEWEES 10-7-9). Similarly, the Interministerial Committee of Economic Planning (CIPE) did not create a new venue for the coordination of R&I policy choices. This committee mirrored the political dynamics characterizing the cabinet in office, representing only a formal coordination table among ministers (INTERVIEWEES 11-3-5).

[6] As argued by one of our experts: "with the CGI they didn't create coherence among the organizations involved, but rather competition between ministers and the Investment for the Future Plan" (INTERVIEWEE 15).

In the Italian case, an additional layer of complexity was added by the important contributions that different sectorial ministers (e.g., the Ministry of Health, the Ministry of Agriculture and Environment), played in the evolution of national R&I strategies. Each of them supervises a specific public research organization like the National Institute of Health (*Istututo Superiore di Sanità*) and the Council for Agricultural Research and Analysis of Agricultural Economy. Nonetheless, as described in Fig. 6.5,[7] the primary funder remains the Ministry of Higher Education and Research. This ministry is by far the main financial contributor of national R&I policy efforts, followed by the Ministry of Economic Development and the Ministry of Health. Specifically, the latter is an important sectorial player in national R&I policy. It manages its own research funding and it also takes part in several international initiatives and infrastructures (Potì and Reale 2011, p. 33).

The alternation of power between sectorial ministers is once again symptomatic of the fragmentation of the Italian national R&I strategy (INTERVIEWEE 10). Indeed, in contrast to the French case, no

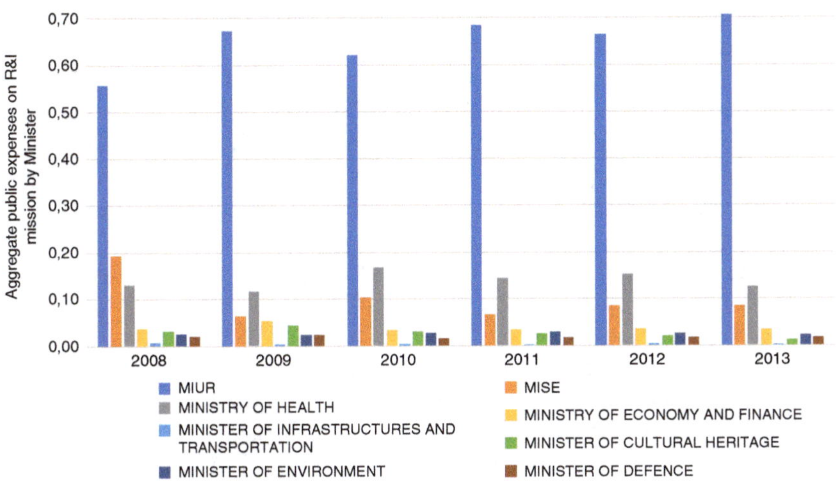

Fig. 6.5 Aggregate public expenses on R&I mission by minister

[7] Data elaborated by the author. Aggregate public expenses by administration and mission (research and innovation mission); source: https://bdap-opendata.mef.gov.it/content/rendiconto-pubblicato-serie-storica-spese-aggregato-amministrazione-e-missione-0.

common budgeting procedure was planned and the agreements over the National Research Plan concerned only broad goals and did not provide any direction regarding the effective implementation of different measures. Therefore, each ministry could essentially carve out a niche for its own interventions and when the areas overlap, coordination could be found internally between specifically concerned ministers (INTERVIEWEE 5). The lack of financial commitment to the multiannual strategies defined in the document has often resulted in unrealistic targets (Nascia and Pianta 2013, p. 24) and unsatisfactory implementation (Grilli and Mariotti 2006).

France and Italy pursued similar coordination strategies which didn't implement any binding obligation for the ministers involved in R&I policy design. In both cases, this resulted in a merely formal primacy of the minister of higher education and research, which nonetheless left enough space for autonomous decisions to the other sectorial ministries involved. Therefore, our expectations regarding the similarity of coordination practices adopted by countries undertaking analogous ministerial coordination practices were confirmed. Both countries have tried to re-centralize their R&I policy coordination powers. Italy attempted to do so by bringing the CIPE structures under the prime minister's office. And France adopted a similar strategy by adding a new layer to the complex national governance of the system with the creation of the General Commissariat for Investments. The coordination practices discussed in Table 6.2 all share a common feature: they do not have their own organizational identity, and they tend to be an arena for exchanging information dependent upon the political will of individual ministers (Braun 2008), rather than an effective policymaking laboratory. Consequently, coordination among ministers is (once again) left to the internal dynamics taking place within cabinets, and to the political attitudes among ministers (INTERVIEWEE 9).

The coordination game between ministers involved in the governance of national R&I policies is complex, and the boundaries of their institutional competences are sometimes fuzzy. Consequently, much of the national R&I is still dependent upon the individuals in charge of the ministries concerned (European Commission 2003). When some coordination was reached (e.g., *Industria 2015* for the Italian case, the Research and Innovation Act and the Innovation plan in France[8]), this was due to either the entrepreneurial capacity of a specific minister or to the collaborative attitude between individuals in charge of different ministries. These

[8] As confirmed by our interviewees (INTERVIEWEES 8-9-10-15).

dynamics are symptomatic of a failure in the organization of the national R&I governance arrangement. The ability to design R&I policies is left in the hands of those with entrepreneurial attitudes, and the individual willingness toward collaboration of each minister; the architecture of the system is not solid enough to function autonomously (INTERVIEWEES 9-10-17-21).

For the Italian R&I this resulted in the fragmentation and duplication of national R&I policy mix. When a set of very similar, and poorly defined, policy competences is established among ministers of an equal institutional weight, it becomes difficult for one minister to take the lead. This will likely increase the tendency to work autonomously, hence intensifying the risk of adopting very similar instruments. The systematic institutionalization of these dynamics can further hamper the policymaking capacity to design long-term national strategies as well as the ability to prioritize specific R&I issues (or missions) in the design of a national R&I policy strategy. Conversely, France attempted to overcome the impasse of R&I ministerial competence divisions by re-centralizing competences into a new institution—the General Commissariat for Investments—which became the pivotal actor in the management of the Investment for the Future plan. This allowed for the re-centralization of the identification of national priorities for the development of R&I strategies, without creating any effective and organic link with current ministerial strategies (OECD 2014). As Musselin (2017) argues, the increasing relevance of the Commissariat hollowed out many of the policy responsibilities previously attributed to the minister for higher education and research, which was increasingly marginalized. This approach shares many similarities with the strategy of layering for equilibrium (Mahoney and Thelen 2010) generally pursued by the French case in R&I policy. Indeed, the CGI was introduced "on top of" the complex institutional system of interaction between ministers and R&I performers, which was smartly circumvented through the leadership role of the new committee. Both countries experienced coordination challenges in the framings of R&I policy action across ministers with joint responsibilities. The organization of R&I institutional layouts requires striking a balance between coordination (e.g., designing an all-encompassing policy strategy) and specialization (e.g., the calibration of instruments to specific contexts and recipients). However, due to the different systems of competence division and related mechanisms of habituation and socialization within organizational structures (March, Olsen, 2005; Merton 1938), there is no best organizational strategy (Pelkonen et al. 2008).

6.4 The Configurations of Interest Representation

Policy design doesn't terminate with the enactment of policy decisions, but it rather continues along the structure of political relations that unfold after instrument implementation (Lascoumes and Le Gales 2007). In this context, governments are increasingly dependent upon the action, acquiescence, and support of others whom they do not directly control (Bressers and O'toole 1998; Goetz 2008; Mazzucato 2013; Nauwelaers and Wintjes 2008; Salamon 2002). Consequently, policy preferences of target populations (policy recipients) also become important, to the extent that the lack of stakeholder support, or legitimacy, for the behavioral change sought by instruments can hamper cooperation and undermine successful implementation (Curley et al. 2020; Dermont et al. 2017; Gross 2007; Ingold et al. 2018; Kammermann and Ingold 2019; Varone and Aebischer 2001). This becomes especially relevant for policy sectors requiring technical competences and skills decision-makers cannot provide by themselves, like in the case of knowledge or technologically intensive activities, where policymakers are likely to rely on the collaboration of external stakeholders for instrument implementation (Guston 1996). This provides R&I performers with the power to shape the final instrument mix and to steer the policy design process toward their expected benefits. While, on the other hand, a portion of this population can miss the opportunity to see their interests represented because of their incapacity to identify shared needs and behave as a political constituency.

To understand R&I policy design, in addition to the internal coordination ability of R&I policymakers, it is also necessary to consider their capacity to interact with the requests of policy recipients. To put it another way, this means that policy takers also matter in the policy design processes but, given their heterogeneity, we still know little about their strategic behaviors. Therefore, we expected that different combinations between R&I performers and their ability to collectively mobilize their resources, and the institutional fragmentation characterizing R&I policymaking in our two case studies, would impact the characteristics of the resulting R&I mix in different ways. Specifically, the higher the capacity of the target population to nullify, or alter, the policy process and the more severe the institutional fragmentation is, the more likely it will be to select a more automatic instrument mix, which leaves greater R&I performers' power in the management of instruments' activation (more automatic delivery structure). Both our cases share a highly fragmented R&I institutional structure, while the landscape of policy takers is quite different across the two countries.

France is characterized by powerful R&I stakeholders who can mobilize their interests and act as a solid constituency, especially in the public research sector. The national higher education and research systems have a mixed nature. The former is defined as a dual tertiary education system (Veltz 2007), due to the separation between highly selective *Grandes Écoles*, mainly oriented toward the training of the future French élites, and universities. The public research system is also split in two, with public research organizations, mainly devoted to research activities, and universities combining teaching and research activities (Musselin and Paradeise 2009). The powerful role of PROs is due to their legacy as pillars of the publicly led Grand Programs, which used to absorb most of the national R&I strategy in the 1950s–1980s. Historically, these organizations used to be heavily supported by the central state and were focused on thematic missions, which happened to be closely related to core national areas of industrial specialization. After the disappearance of what has been a central model of public intervention, the state has inherited a large landscape of organizations with various legal statuses and fragmented competences (Mustar and Larédo 2002; Théry 2004). This delineates a landscape of public research stakeholders in which, on the one hand, there are universities organized in smaller units dispersed across the territory. On the other hand, there are PROs benefitting from a consolidated national network of laboratories (INTERVIEWEES 21-17). Instruments like the ones implemented with the Law on the Autonomy and Responsibilities of Universities, in combination with the Research Act and the creation of the National Research Agency, attempted to place universities at the center of the system as main operators of public research (Musselin 2017). However, a combination of institutional rigidities and stakeholder activism against these changes, mainly form PROs, made this policy change only partially effective and short-lived (Lepori et al. 2017). This cleavage is historically grounded in the national R&I system because "*[…] the executive élites are not trained within universities*" but in the *Grandes Écoles*.[9]

French policymakers have an extensive record of consultations with these actors, probably as a reaction to their strong activism. The first event dates to 1956, with the *Colloque de Caen*, followed by *Les Assisses Nationales* (1981), which anticipated the adoption of the 1982 Law on Research. There have also been consultations during the design of the 1999 Innovation Plan and the 2003 Innovation Act, after the creation of the SLR movement, as a reaction of the research world to the first drafts of the public research reform (2003). A second consultation with R&I performers was organized in 2004, the *General State of Research* (Ministère de l'Education nationale de l'Enseignement supérieur et de la Recherche 2004). Similarly, another consultation took place during the definition of the National Strategy for Research and Innovation (2009), then again in 2010 with the Roundtable for Industry and finally in 2013 with the National Conference on Higher Education (*Assises de l'Einsegnement Superieur*). These events have represented consultative, rather than strategic, forums for discussions in which the establishment of an effective collaboration often seemed impossible (INTERVIEWEE 19). However, the opportunities for dialogue have nevertheless portrayed a decision-maker's attitude oriented toward creating a more favorable environment for building consensus among R&I performers (OECD 2014).

On the side of private R&I performers, the dichotomy between big enterprises versus SMEs has been a constant in the national governance of the sector. The significant legacy of state-led R&I programs, which mainly relied on large national enterprises, has heavily influenced the evolution of national strategies. In parallel with the beginning of the liberalization of the French national economy, there has also been a shift in attention

[9] As argued by one of the experts we interviewed (INTERVIEWEE 13): "*it is necessary to understand that, in France, universities didn't use to be relevant institutions. Differently from other countries, the executive élites are not trained within universities. This is historically related to the French Revolution. At that time, universities were related to the church. Since the revolution was anticlerical, we have created new élites detached from the church, with the creation of the Grandes Écoles; which has been enhanced during the Napoleonic era, and in 1945 with the creation of the École Nationale d'Administration (ENA, National School of Administration). The French bourgeoisie trains their children in the Grandes Écoles, not at universities. In these institutions the goal is not to 'learn to do' but to 'learn to learn,' therefore French élites tend to be quite detached form research and science (except for some highly research-intensive Grandes Écoles institutions* e.g., *École Normale Supérieure or École Polytechnique).*"

toward the role of SMEs, especially boosted by the creation of competitive clusters. However, such a focus toward SMEs has slowly decreased across the years, as discussed in Chap. 4 with the description of the Investment for the Future program. This trend should also be read in the light of different R&I performers' capacity to mobilize their interests. In this context, business demands were articulated both formally—through consultation bodies like the Permanent Commission for Consultation with Industry (CPCI), advising the Ministry on Economic Affairs[10]—and informally via ad hoc consultative bodies, often chaired by representative of the industrial world, drafting reports on specific issues on the behalf of the prime minister. One instance of this process was the Blanc Report, initiated by Christian Blanc (see Sect. 4.4.1), which triggered the creation of the competitive clusters (Schoen et al. 2008, p. 22). Other informal channels of communication can take place on an issue-based mechanism between the minister with delegated competences in industry and the MEDEF. As discussed in Chap. 4, the MEDEF is the largest employer's federation in France, which primarily represents SMEs. Conversely, big enterprises tend to run their businesses autonomously; they can rely on the lobbying of the MEDEF as a forerunner, but they can also interact personally with the interested minister (INTERVIEWEE 20). This practice should not be interpreted as a standard strategy for interaction, but it should rather be read as an action by exception, which can be activated by the political powers at the highest level when they choose to interfere (INTERVIEWEE 12). The possibility for big enterprises to have a voice in the policymaking process was also dependent upon the political attitudes of elected politicians, more specifically on their openness toward the business world. The history of the tax credits for investments in research, presented in Chap. 4, might indeed have been shaped by these dynamics to the extent that, in the contemporary literature, we also find many references to the cooperative attitude of the former President of the Republic (and minister of economy finance and industry), Nicolas Sarkozy, toward the industrial world (Cole et al. 2008). Therefore, the possibility to overcome formal channels of interaction is said to be dependent upon the political permeability of policymakers toward the business world.

The Italian R&I system is also characterized by a dichotomy in the public research sector. On the one side, there are PROs exclusively involved

[10] This brings together experts from this and other ministries, industry representatives from the enterprise association MEDEF, and other stakeholders.

in research activities, and on the other side, there are universities, combining education and research activities. The National Council for Research (CNR), the main PROs in the country, used to have an important role in the coordination and funding support for national research policies. It used to play a double role in the national research landscape, being both a research funder and a research performer. However, the relevance of this organization was slowly transformed by hollowing out its funding competences, as described in Chap. 5 (see Sect. 5.3.2). By means of small reforms throughout the early 2000s, the landscape of public research organization was re-organized, and the main national PRO, a quasi-research funding agency, was transformed into a research performer, competing with universities in the race for research funding. Here, we can draw some parallels with the French experience and the attempt to decrease the power of PROs via the Research Act and the reform of higher education. In this case, the combination between the high capacity to mobilize interest from public research stakeholders, as well as the clear attack on their power, resulted in a substantial conservation of the status quo. Differently, a combination between a low capacity for mobilizing the interests of the Italian public research performers, together with a more gradual, and less immediate, alteration of the governance of the sector resulted in a policy change.

In the Italian case, formal consultations with key R&I stakeholders are performed on a regular basis; however, their involvement is still partial and has only limited influence on policymaking (European Commission 2006). There are limited institutionalized opportunities for R&I performers to raise their voices in the policymaking process, and when it is strictly needed these practices are kept at the institutional level. Consequently, the combination between the inability of national policymakers to coordinate different actors, as well as the lack of a long-term political vision for national R&I strategies, can pave the way for more structured stakeholders (e.g., resourceful business actors or powerful research centers) to promote their specific interests. Contrary to the French case, the Italian R&I system is characterized by an endemic fragmentation in the representation of R&I interests on both the public and the private side. On the first side, universities and PROs do not manage to represent a stable interlocutor of the government in R&I policymaking (INTERVIEWEES 8-4), and, despite some important national research centers, these stakeholders are rather unsuccessful in collectively representing their interests. The landscape on the side of industrial and business actors seems to reflect a similar endemic fragmentation but also greater similarities between the two cases. The representation of the industrial world hinges upon the General Confederation of Italian Industry (Confindustria), an association with voluntary membership representing companies, their values, and interests, in national policymaking. Nowadays this organization is facing important challenges in terms of representativeness (INTERVIEWEES 8–11). Indeed, there is a widespread perception regarding the increasing incapacity of Confindustria to represent the requests of big enterprises in the field of R&I, which has been somehow translated into a greater attention toward the needs of SMEs and increased autonomy of bigger enterprises in their innovation-related investments. All of that is also reflected by the fact that when designing policy, in addition to the dialogue with Confindustria, ministers also organize one-to-one meetings with selected industrial counterparts (INTERVIEWEE 5). This practice recalls the French case, where only SMEs congregate into networks of enterprises to represent their interests across the policymaking process. Meanwhile, large enterprises act autonomously through a set of different informal strategies. This suggests how the two cases share some similarities in the way private R&I interests are organized, while they experience different configurations of interest representation in the public research sector.

6.4.1 *Intermediary Agency Versus Ministerial Intermediation*

As discussed in Chap. 3, to institutionalize conflict and to overcome uncertainties in their dialogue with R&I target populations, policymakers have developed various strategies for intermediation (Braun 1993; Van der Meulen 1998). In the French R&I governance, the task of translating political guidelines and thematic priorities into different practices involves specific intermediary organizations, like public agencies. These actors are located between societal spheres, with the job of institutionalizing social interfaces and creating common contexts for action (Hartmann and Kjaer 2015). Their development and diffusion reflect New Public Management (NPM) principles, according to which national bureaucracies are increasingly relying on external bodies and committees to formulate and implement their policies (OECD 2005). These can acquire various organizational characteristics, according to the existing structures of interest intermediation between policymakers and target groups. An important instance of this strategy was the creation of the National Research Agency with the 2006 Research Act. This institution was granted administrative and financial decision-making autonomy, and it was placed under the supervision of the minister of higher education and research.[11] Its mission was to fund public research performers, according to the thematic priorities identified by the government. However, concerns from the research community regarding the government's powers to establish thematic priorities for the national public research strategy paved the way for some changes in its internal governance structure, with the introduction of the "Alliances"[12] in 2010. This is a successful instance of stakeholders' interest mobilization which modified the underlining nature of the agency. Indeed, despite an initial inflow of funds and the organizational flexibility provided in the design of thematic priorities, the ANR struggled to find its position in the French R&I governance, which remained sensibly weighed toward PROs (Lepori et al. 2017; OECD 2014). Another public agency was operating since 1967 in the field of the valorization of industrial research and public support to SMEs: the National Agency for the Promotion of Research (ANVAR—*Agence Nationale de Valorisation de la Recherche*). Its activities were combined with the creation of a specialized funding institution (SOFARIS), which managed a guarantee fund for the development of SMEs, the so-called Deposits and Consignments Fund (CDC).[13] In 2005, almost in co-occurrence with the creation of competitive clusters, these agencies were merged into a unique organization—OSEO—a holding company owned by the State, mainly reporting to the Ministry of Economy and Finance. This agency rationalized the national structure of innovation support for SMEs, and it became the main operator in charge of managing and implementing policies in this field (European Commission 2005). Its nature has been evolving over time, first with the absorption of the Industrial Innovation Agency (2008) after being itself absorbed by BPI

[11] The agency took over the support actions previously financed through the FRT and FNS; subsidizing basic and applied research undertaken by public research organizations, universities, and SMEs—as well as some direct institutional funding to research laboratories (Arnold 2007; Lepori et al. 2017).

[12] These were coordination institutions bringing together different public stakeholders in a given research domain to support coordination in the programming of the agency. These coordination groups brought together public research stakeholders, according to their research domain, to enhance cooperation in setting the programmatic priorities of the National Research Agency (Zaparucha 2010). Basically, the new role of Alliances entailed reinstating the planning function itself, within the PROs, which ran counter to the previous trend of separating powers implemented by the Research and Innovation Act (OECD 2014).

[13] The Caisse des Dépôts et Consignations (Deposits and Consignments Fund) is a state bank which funds companies and is heavily involved in innovation and SMEs financing (OECD 2014).

France (a public investment bank funding innovation activities) in 2013. These are the two most important instances of French strategies for interest intermediation, which worked with public research performers and the business sector, respectively.

Conversely, the design of Italian R&I policies tends to be mainly performed at the governmental level. Italy doesn't envisage any specialized agency in the field of R&I, and the intermediation between policymakers and policy takers' interests takes place at the ministerial level. National R&I policies are steered directly by different ministers with R&I competences, under the supervision of the parliament. There have been some attempts to establish specific actors with similar competences (like *Sviluppo Italia* and the National Agency for the Diffusion of Technologies for Innovation), but their role has always been marginal, especially because of their short life. Overall, there is a lower tendency toward creating dialogue between ministers and R&I performers, and if that is strictly needed, these practices are kept at the institutional level (INTERVIEWEE 9). To some extent this is also related to the fact that policymakers are no longer solicited by social partners or business representatives to do so. Indeed, contrary to their French counterparts, Italian R&I stakeholders are less capable of mobilizing their interests during policy design.

In the public research sectors, there used to be a quasi-agency actor, the National Research Council (CNR), which played a double function as both research funder and performer. This organization managed the administration and selection of national research projects by funding block-grants directly to individual researchers while also carrying out its own research activities by means of strategic projects. However, after going through a profound reorganization process, its coordination and funding functions were progressively moved to the minister for higher education and research. Despite contrasting views (also among our interviewees) regarding the titularity and the effectiveness of CNR funding activities, there is widespread agreement concerning the empty space left by this reform which didn't account for any other actor able to represent a unified center of political responsibility to whom stakeholders from the public research sectors could raise their claims and suggestions (INTERVIEWEES 7-10-9). The inability of coordinating different actors, and the lack of an intermediary organization that can institutionalize social interfaces and create common contexts for action, make policymakers prisoners of actors supporting their specific interests—like some entities from the business world or some powerful research centers and universities

(INTERVIEWEE 4). Consequently, the design of national policy strategies is highly influenced by political volatility and by the frequent cabinet reshuffles characterizing the national political system.

Overall, the structure of interest intermediation between decision-makers and R&I performers is quite fragmented in both countries. Italian R&I performers suffer from a lower capacity to influence the policy process. This is matched with a direct intermediation strategy, which bears many political uncertainties, where policymakers seem to be loosely attracted by R&I themes—both for their own political ambitions and as a strategy for national economic development. Conversely, R&I performers in France have shown a greater capacity to influence the policy process, despite the landscape, on both the private and public sector, looking highly disaggregated. Interest intermediation is organized through specialized public organizations, which are differentiated according to their users. In both cases, interest representation of private R&I performers is structured according to specific organizations, although the overall decision-making system tends to be biased toward the direct intermediation strategies adopted for dialoguing with big enterprises. These actors have bargaining power by exception, according to the different degrees of permeability of elected politicians toward the business world.

We expected that the higher the capacity of the target population to nullify or alter the policy process, and the more severe the institutional fragmentation is, the more likely it will be to select a more automatic instrument mix, which leaves greater powers to R&I performers in the management of their activation (more automatic delivery structure). Therefore, given that both countries share an external ministerial organization strategy, we expected that the characteristics of R&I performers could influence the resulting instrument choices differently.

Both countries show a high share preference for low delivery types of instruments, which is respectively 68% of the total delivery choices for France and 46% for Italy. In the first case, contrary to our expectations, governments have tended to re-centralize policy instruments' implementations in their hands as a reaction to the great powers of the target population to nullify or alter the policy process. After the creation of public agencies like BPI France and the ANR, policymakers have consistently modified their approach toward R&I funding. They are increasingly converging toward a "public holding model" (Eparvier et al. 2011) where, through the provision of different funds, the central administration retains financial stakes in innovation organizations, without the costs associated with their management. Contrary to our expectations, Italy, a country characterized by high fragmentation in R&I stakeholders' representation, sees a greater share of choices for automatic instruments (33%) compared to France (11%). This can be explained by the increasing collaboration with private banks for the financing of innovation. At the beginning of the 2000s, specifically with the Berlusconi II cabinet, private banks provided finance to enterprises (mainly SMEs), and public actors participated by backing up with public guarantees loans or cash advances to business actors. In this pattern, instead of behaving as shareholders, policymakers became the bearers of the risks related to R&I investments, but they were cut off from future benefits produced by the firms they supported. Basically, the private banking system provided available credits to R&I performers, overcoming the sluggishness related to the public management of these funds, while policymakers were bearing the risks of possible associated failures. The use of a credit system based on banks eased the planning of resources for investments in innovation by enterprises, but it didn't express any specific policy goal in terms of R&I strategy (Gallo and Silva 2006).

This suggests how, in the French case, as opposed to what was hypothesized, the activism of national R&I performers has been contrasted with a greater centralization of powers in the delivery of policy instruments (Fig. 6.6). Despite the communicative attitude of policymakers toward

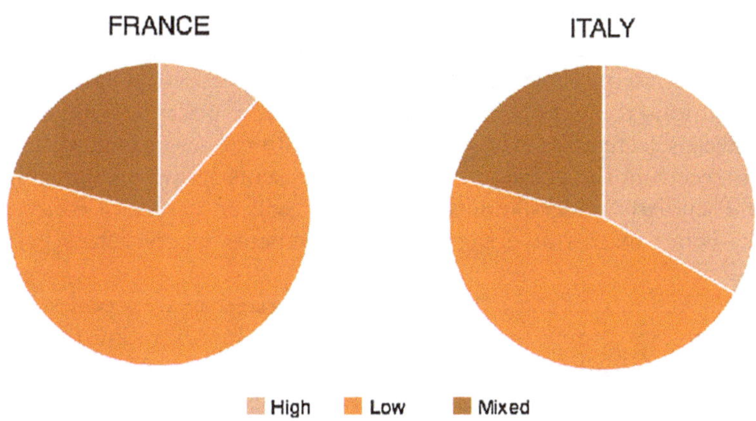

Fig. 6.6 Choices of instrument delivery component by country

R&I performers, as demonstrated by the opportunities for dialogue with R&I stakeholders discussed in previous paragraphs, the government strategy has heavily relied on a centralization of powers. This happened both through the creation of public research organizations and through the re-centralization of policymaking authority in the hands of the prime ministers, during the Investment for the Future plan. While in the Italian case, the combination between a fragmentation of R&I performers' interest representation strategies and the incapacity to design long-term ministerial strategies of interaction resulted in an increasing reliance on banks for the funding of innovation. This strategy unburdened ministers from their duties related to the implementation of funding instruments but removed some powers of control from policymakers concerning the choices over funding public guarantees loans or cash advances to business actors. Therefore, the reliance on automatic delivery structures is needed because of the competences and resources the private credit system can mobilize and the public sector miss out on.

6.5 Final Remarks

In this research, we are interested in understanding the political dynamics that stand behind policy instrument selection in R&I. We explored how the interactions between actors involved in the governance of the sector, including knowledge producers and developers, public and private actors, together with policymakers, influenced this process. The analysis illustrates the way governments design more ambitious policy mixes aimed at solving the complex, and cross-sectorial in nature, social challenges contemporary societies are facing. The research provides both theoretical and empirical contributions to support an explanation on how policy instruments are effectively selected from the perspective of both the actors making and the actors receiving the decisions. This also represents an interesting addition to the scholarly understanding of how the holistic nature of many R&I policy issues maps out governance arrangements where actors become interdependent.

To make sense of these dynamics, we theorized an R&I policy instrument typology (family-shape-delivery) capable of interpreting the behavioral changes sought by different instruments. First, focusing on national decision-makers, we investigated the role different partisan political orientations had in shaping their choices for alternative policy instrument characteristics. We know that policy instruments are chosen with a purpose

(Edler et al. 2016), so the way these are designed and implemented represents a manifestation of different political views. Therefore, we expected that specific R&I instrument blends would be associated with different political coalitions. Results show how instrument choices tend to be slightly more polarized in Italy, compared to France, where it is possible to identify clearer bipartisan preferences across instrument characteristics. This suggests how our expectations about the association between R&I instrument blends and different political coalitions in office were not confirmed. We can conclude that these choices are only an extremely moderate manifestation of different political orientations of the cabinet in office. Since France and Italy are characterized by different traditions of R&I policy types, respectively mission-oriented and diffusion-oriented, we expected to identify different trends in the distribution of government choices for policy instrument features. Indeed, despite the two countries having followed a binary path of evolution in their policy choices, when zooming in on the specific dynamics characterizing these developments, their constitutive elements seem more evident. France shows clearer preferences for expenditure instrument families, higher coercive shapes, and low delivery structures that rely on existing bureaucratic structures for the implementation of public action. All features that are in line with the national traditions of R&I mission-oriented policy types. On the other hand, Italy shows a greater share of preferences for fiscal instruments, lower coercive shapes, and higher automatic delivery components. This pattern of choices is highly consistent with a diffusion-oriented policy type whose instrumentation relies on less authoritative and more automatic instrument mixes. In this context, policy decisions tend to be taken in a decentralized manner, and the implementation of these tasks is delegated to either industrial associations or cooperative research organizations (Ergas 1987).

Further, by narrowing the focus on those actors involved in the policymaking of national R&I strategies, we investigated the way different morphologies of ministerial organization practices, and their related coordination challenges, influenced policymaking activities. Since the success of this institutional structure is highly dependent on the ability to create dialogue among different policy responsibilities, we expected that countries undertaking similar ministerial specialization strategies would likely adopt similar coordination practices (informal institution). Our expectations were confirmed: both of our cases have tried to re-centralize their R&I policy coordination powers. Italy attempted to do so by bringing the CIPE under the prime minister's office. And France adopted a similar strategy by adding a new layer to the complex national governance of the system with the creation of the General Commissariat for Investment. Both countries' coordination practices (Table 6.2) share a common feature: they do not have their own organizational identity, and they tend to be an arena for exchanging information dependent upon the political will of individual ministers (Braun 2008), rather than an effective policymaking laboratory. Coordination among ministers is once again left to the internal dynamics taking place within cabinets and to the political attitudes between ministers. Much of the national R&I policymaking capacity is still dependent upon the entrepreneurial capacity to establish practices for coordination of the individuals in charge of the ministries concerned (European Commission 2003).

Finally, we focused on the interactions between policymakers and policy takers, investigating the way interest intermediation effectively takes place and the extent to which this influences R&I policy design choices. We expected that different combinations of R&I performers' ability to collectively mobilize their resources and ministerial institutional fragmentation would impact the characteristics of the resulting R&I mix in different ways. Specifically, we hypothesized that the higher the capacity of the target population to nullify or alter the policy process, and the more severe the institutional fragmentation is, the more likely it will be to select a more automatic instrument mix, which leaves a greater share of R&I performers' power in the management of their activation (more automatic delivery structure). Both our cases share a highly fragmented R&I institutional structure; however, the national landscape of R&I policy takers is quite heterogenous across countries. France is characterized by powerful R&I stakeholders who can mobilize their interests and act as a solid constituency, especially in the public research sector. As a reaction to the great powers of the target population to nullify and alter the policy process, decision-makers have tended to re-centralize policy instruments' implementations in their hands. After the creation of public agencies like BPI France and the ANR, policymakers have consistently modified their approach toward R&I funding. They are increasingly converging toward a "public holding model" (Eparvier et al. 2011) where, through the provision of different funds, the central administration detains financial stakes in innovation organizations without the costs associated with their management. Differently, Italy is characterized by an endemic fragmentation in the representation of R&I interests on both the public and the private side. Compared to the French case, the country sees a greater share of choices for automatic delivery components. This can be explained by the increasing collaboration with private banks for the financing of innovation. Basically, the private banking system can provide available credits to R&I performers, overcoming the sluggishness related to the public management of these funds, while policymakers bear the risks of the possible related failures. This strategy eased the planning of resources for investments in innovation by enterprises, but it didn't express any specific policy goal in terms of R&I strategy (Gallo and Silva 2006). The main outcome was that ministers were relieved of their responsibilities for implementing funding instruments, but it also resulted in policymakers losing some control over decisions regarding funding public guarantees, loans, and cash advances to businesses (which were in the bank's hands). Consequently, automatic delivery structures become necessary because only the private credit system has the competencies and the resources that the public sector needs.

As discussed throughout the comparative case study of France and Italy, these two countries share many similarities in their diachronic development of national R&I governance. However, as demonstrated by the different patterns of R&I instrument choices (as described in Sect. 6.2), this can also harbor substantial internal differences in the relationships between policymakers and the target population. Both cases have the necessary R&I instrumentation at their disposals, but they are not equally able to make use of them to their advantage.

First, these differences can be seen in the political saliency that these policy issues acquire across the two cases. R&I policy choices represent a special investment in long-term intangible assets that will (likely) generate profits in the future. This is different from regular investments in tangible assets, such as capital expenditures, because of its longer time horizon and higher risk (Bhattacharya et al. 2017, p. 2). Consequently, when making these choices a political strategy is needed because the investment made today will produce results tomorrow. Indeed, if policymakers do not construct a vision—some expectations—regarding what they want to achieve through R&I policy, they might fall in the trap of creating structures in which bureaucracies derive their *raison d'être* only from the administration of subsidies, leaving apart their willingness to pursue a broader national strategy for development. Similarly, R&I performers will also struggle to foster their businesses, given the high volatility related to the lack of political leadership. For the Italian case, the generally low political salience of these issues is evident. R&I has never been a priority for Italian governments, and it has never been central in their decision-making agenda (INTERVIEWEES 7-10-11). Indeed, the Italian cultural landscape has historically displayed a widespread negative outlook toward science and technology (Nuvolari and Vasta 2015, p. 282). Moreover, another important limitation characterizing the Italian politics of R&I refers to the tendency of the system to wipe out what previous governments have done. Newly elected governments find it hard to explain to their own majority that there is no need to go back to square one every mandate and that what is needed is simply to forge ahead, to adjust what has been previously done (Bassanini 2009). This creates further fragmentation in the evolution of R&I policy strategies. Instrument choices tend to be isolated from a broader systemic view, and R&I policies seem to be mainly designed through one-off interventions approved in the framework of yearly budgetary laws. This pattern doesn't only create redundancies, which sometimes lead to the overlapping of similar instrument shapes, but it also creates confusion in the target population which tends to be lost in the jungle of very similar tools at their disposal (INTERVIEWEE 10). Such a fragmentation and duplication of instruments is matched with a scarcely effective monitoring and the uncertainties related with the implementation process and availability of resources (Potì and Reale 2011). The Italian case suffers from an endemic fragmentation and duplication of instruments, and actors tend to follow a short-term logic of layering for change through the design of one-off interventions. This characterizes a system of national development strategy, in which "everything must change to remain the same" (Dosi and Roventini 2022, pp. 270–271). The evolution of the national R&I instrumentation doesn't seem to follow any clearly identified long-term strategy, embedding a high degree of misinformation between those who allocated the funds and the beneficiaries (De Maio 2011).

In the French innovation system, despite some degree of fragmentation of R&I national strategies, it is possible to trace an evolution, over time, toward the design of long-term national R&I missions. There is widespread agreement about the political willingness to maintain a role for the public in strategic industrial sectors of the national economy (INTERVIEWEE 17). Policy instruments re-call each other, or at least are attempting to go in that direction. Despite the siloed administration of R&I policies, the instruments implemented in one sector are complementary to the actions undertaken in another sector. An instance of this is the evolution of the tax credit for research. The reform in 2004 took place at the same time as the implementation of the competitive clusters, and it was somehow meant to simplify the usage of the credit by SMEs, which were among the main pillars of the cluster's development. The second reform of the tax credit (2006) included newly eligible expenses related to the subcontracting of research made by PROs, universities, and other research centers. This happened in parallel with the creation of the new tool for grouping HEIs and PROs in geographical proximity (PRES), created by the Research Pact (2006), which were meant to become the main beneficiaries of the new subcontracting research services (also included in the credit deductions).

These results substantiate the theoretical and empirical capacity of our instrument typology to provide insights into the constitutive differences embedded in instruments families, shapes, and delivery structures. This demonstrates how it is not enough to look at the types of instruments selected (families), but it is necessary to understand their capacity to induce a specific behavior while addressing the right target. Policy instruments containing similar coercive principles can be applied in different ways (Vedung 1998); in addition to social control, each tool embeds a particular way of exercising it (Kassim and Le Galès 2010; Lascoumes and Le Gales 2007). The choice, and the particular way in which instruments are structured and assembled, reflects the interaction between decision-makers and the political clientele of the policy. This allows us to explore the extent to which decision-makers accounted for target groups' stakes and interests along the policy design process (Skodvin et al. 2010), and how differently powerful actors managed to exploit institutional organization structures to influence the final shapes of policy mix evolutions.

Policy instrument selection and design into different packages is not a rational, linear, or purely technical exercise. It rather represents a *muddling through* (Lindblom 1959), an inherently political process involving different actors, interests, and constraints (Linder and Peters 1989; Peters 2000). For all these reasons, we believe that in the study of policy design it is important to overcome the restricted focus on policy instruments as means through which governments use their legitimate power to shape public action (instrument families). Exercising power means to obtain collaboration—that is, obtaining behavior in line with our expectations (Stoppino 2001). Consequently, to understand how authority is effectively exercised, we need to make sense of the different relationships policy instruments (or combinations of thereof) establish between policymakers and the target population. The proposed policy instrument typology (family-shape-delivery) can provide a deeper understanding of the behavioral and political characteristics embedded in instrument action while showing how both the choice and the particular way instruments are crafted reflect different types of interactions between decision-makers and policy recipients.

6.6 Policy Implications, Limitations of the Study, and Venues for Future Research

Throughout the research investigation process foundational to this book, we observed two important phenomena describing the relationship between scholarly theorizations and the application of this knowledge for policymaking.

The first refers to the perceptions over the mechanisms driving innovation. Both of our case studies adopt a consolidated linear view of the innovation process, which appears to be dominant among many policymakers and innovation stakeholders. This model postulates that innovation starts with basic research, is followed by applied research and development, and ends with production and diffusion (Godin 2006, p. 639). Such an interpretation of innovation is often contraposed to the evolutionary and institutional approach, which sees innovation as a socially embedded process where the dysfunctionalities that may negatively affect innovation activities are context-dependent; hence, market failure is one among other possible failures (Borrás and Edquist 2019; Chaminade and Edquist 2006). Our analysis has adopted an actor perspective to understand the capacity of different governments to get things done. Therefore, during the reconstruction of the policy process (through interviews and document analysis), it emerged how policymakers tend to make use of the narratives related to the latest developments in the field of innovation research, like system-based innovation approaches. However, once they have to agree upon how to coordinate their action internally (e.g., among ministers) and externally (e.g., in their interaction with stakeholders), they tend to adopt interventions consolidated in the linear view of innovation.

The second important empirical gap we identified relates to policy instruments. Especially throughout the interviews, it became apparent how the heuristic of policy instrument was a merely theoretical concept for some interviewees. This can explain why, in many national policy design contexts, policymakers rarely have a clear overview of what has been done before and of which interventions have already been put into action. This often leads to an inconsistent layering of interventions and a related low-uptake of these instruments from the recipients, who are lost in the jungle of very similar tools at their disposals. Consequently, we believe that further scholarly attention should be devoted to identifying more accurate, and empirically grounded, classifications of R&I policy instruments. This exercise can provide a more accessible analytical toolkit to the actors shaping national policymaking, and it can also be beneficial for the scholarly community, who will be able to work within a common framework of analysis capable of tracing the activity of instruments over time more easily. We believe that it is important to further stress the necessity to work on an integrated theoretical framework in the study of policy instruments. This, as demonstrated by the results of our analysis of R&I policy, should prompt the investigation of tools and their selection process from a two-way perspective, where both policymakers and recipients are considered as actors able to influence the characteristics of public action. Similarities in the way governments use their legitimate power to shape public action (instrument families) do harbor substantial internal differences in the relationships between policymakers and target populations these can activate, which we have only been able to grasp, thanks to the introduction of instrument shape and delivery structure categories.

To conclude, we want to mention some of the limitations of this study. First, during the empirical analysis, especially in the interview process, we faced some issues regarding the low accessibility of private actors in the data collection process. Our investigation was mainly oriented toward the policy formulation process; therefore, we are confident in stating that by interviewing policymakers and representatives of business organizations we reached the needed data saturation point. Nevertheless, we believe that further research in this field would consistently benefit from a broader spectrum of data coming from private enterprises. Secondly, given the time frame of analysis, and the limitations related with the management of our research, we decided to leave out two industrial policies: namely, *Industrie du Future* (Industry of the future) for the French case and Industria 4.0 (Industry 4.0) for the Italian case. This choice was mainly driven by the fact that these two policies were going through a formulation process during the period of our data collection. Therefore, given the fact that the numbers of years under investigation was in line with the appropriate time frame for analyzing policy change identified by the literature (Sabatier 1986), we decided to leave out these two newly arisen policies. We believe that this decision can pave the way for a (follow-up) comparative analysis of these two policies in light of the analytical framework developed in the present study; this will surely provide a valuable contribution to the existent literature in the field. Finally, we are aware that this study adopted a macro perspective on the analysis of R&I policy sectors. Our aim was precisely to understand the nationwide dynamics shaping the design of these policies in two different countries. Therefore, this paves the way for a new research venue focusing on the regional, or local, R&I perspective that can test the validity of our policy instrument selection framework on a different level of analysis.

Bibliography

Arnold, E. 2007. CREST 3 % OMC Third Cycle Policy Mix Peer Review Country Report France.

Bassanini, F. 2009. Twenty Years of Administrative Reform in Italy. *Review of Economic Conditions in Italy*: (3) 369–391.

Bergek, A., M. Hekkert, S. Jacobsson, J. Markard, B. Sandén, and B. Truffer. 2015. Technological Innovation Systems in Contexts: Conceptualizing Contextual Structures and Interaction Dynamics. *Environmental Innovation and Societal Transitions* 16: 51–64.

Bhattacharya, U., P.-H. Hsu, X. Tian, and Y. Xu. 2017. What Affects Innovation More: Policy or Policy Uncertainty? *Journal of Financial and Quantitative Analysis* 52: 1869–1901.

Binder, S.A., R.A.W. Rhodes, B.A. Rockman, J.G. March, and J.P. Olsen. 2009. Elaborating the New Institutionalism.

Borrás, S., and C. Edquist. 2019. *Holistic Innovation Policy*. Oxford University Press.

Braun, D. 1993. Who Governs Intermediary Agencies? Principal-Agent Relations in Research Policy-Making. *Journal of Public Policy* 13: 135.

———. 2008. Lessons on the Political Coordination of Knowledge and Innovation Policies. *Science and Public Policy* 35: 289–298.

Bressers, H., and L. O'toole. 1998. The Selection of Policy Instruments: A Network-Based Perspective. *Journal of Public Policy* 18: 213.

Capano, G. 2008. Looking for Serendipity: The Problematical Reform of Government within Italy's Universities. *Higher Education* 55: 481.

Capano, G., M. Regini, and M. Turri. 2016. *Changing Governance in Universities: Italian Higher Education in Comparative Perspective*. London: Palgrave Macmillan.

Capano, G., A. Pritoni, and G. Vicentini. 2019. Do policy Instruments Matter? Governments' Choice of Policy Mix and Higher Education Performance in Western Europe. *Journal of Public Policy* 40: 1–27.

Chaminade, Cristina, and Charles Edquist (2006). "From theory to practice: the use of the systems of innovation approach in innovation policy." In J. Hage, & M. Meeus (Eds.), Innovation, Science, and Institutional Change A Research Handbook Oxford University Press : 141–163.

Cole, A., J. Levy, and P. Galès. 2008. From Chirac to Sarkozy: A New France? In *Developments in French Politics 4*, 1–21. Basingstoke/New York: Palgrave Macmillan.

Curley, C., R. Feiock, and K. Xu. 2020. Policy Analysis of Instrument Design: How Policy Design Affects Policy Constituency. *Journal of Comparative Policy Analysis: Research and Practice* 22: 536.

Cytermann, J.-R. 2006. L'architecture de la loi organique relative aux lois de finances (LOLF) dans les domaines de l'éducation et de la recherche: choix politiques ou choix techniques? *Rev. française d'administration publique* 117: 85–93.

De Maio, A. 2011. *L' innovazione vincente*. Brioschi Editore, Milano.

Dermont, C., K. Ingold, L. Kammermann, and I. Stadelmann-Steffen. 2017. Bringing the Policy Making Perspective in: A Political Science Approach to Social Acceptance. *Energy Policy* 108: 359–368.

Dosi, G., and A. Roventini. 2022. The Leopard: How a Post-Fascist Party Rose to Power in Italy. *Intereconomics* 57: 270–271.

Edler, J., P. Cunningham, A. Gök, and P. Shapira. 2016. *Handbook of Innovation Policy Impact*. Cheltenham: Edward Elgar Publishing.

Edquist, C., and S. Borrás. 2013. The Choice of Innovation Policy Instruments. *Technological Forecasting and Social Change* 80: 1513.

Eparvier, P., O. Mallet, and L. Rivoire. 2011. Erawatch Country Reports 2011: France.

Ergas, H. 1987. The Importance of Technology Policy. In *Economic Policy and Technological Performance*, ed. P. Dasgupta and P. Stoneman. Cambridge: Cambridge University Press.

European Commission. 2003. European Trend Chart on Innovation – Country Report France.

———. 2005. European Trend Chart on Innovation – Annual Innovation Policy Trends and Appraisal Report: France.

———. 2006. European Trend Chart on Innovation Country Report: Italy.

Evangelista, R. 2007. Rilevanza e impatto delle politiche dell'innovazione in Italia. I risultati delle indagini CIS. *Economia e Politica Industriale* 1: 103–124.

Fagerberg, J. 2017. Innovation Policy: Rationalities, Lessons and Challenges. *Journal of Economic Surveys* 31: 497–512.

Foray, D. 2009. *The New Economics of Technology Policy*. Cheltenham: Edward Elgar.

Foray, D., D.C. Mowery, and R.R. Nelson. 2012. Public R&D and Social Challenges: What Lessons from Mission R&D Programs? *Research Policy* 41: 1697–1702.

Gallo, R., and F. Silva. 2006. Le condizioni per crescere: diagnosi e proposte per il sistema produttivo. Il Sole 24 ore.

Godin, B. 2006. The Linear Model of Innovation: The Historical Construction of an Analytical Framework. *Science, Technology & Human Values* 31: 639–667.

Goetz, K.H. 2008. Governance as a Path to Government. *West European Politics* 31: 258–279.

Grilli, L., and S. Mariotti. 2006. Politiche per l'innovazione e cambiamento strutturale in Italia. *L'industria*: 365–396.

Gross, C. 2007. Community Perspectives of Wind Energy in Australia: The Application of a Justice and Community Fairness Framework to Increase Social Acceptance. *Energy Policy* 35: 2727–2736.

Guston, D.H. 1996. Principal-Agent Theory and the Structure of Science Policy. *Science and Public Policy* 23: 229–240.

Hartmann, E., and P.F. Kjaer. 2015. *The Evolution of Intermediary Institutions in Europe*. London: Palgrave Macmillan UK.

Ingold, K., I. Stadelmann-Steffen, and L. Kammermann. 2018. *The Acceptance of Instruments in Instrument Mix Situations: Citizens' Perspective on Swiss Energy Transition*. Vol. 48, 103694. Research Policy.

Kammermann, L., and K. Ingold. 2019. Going Beyond Technocratic and Democratic Principles: Stakeholder Acceptance of Instruments in Swiss Energy Policy. *Policy Sciences* 52: 43–65.

Kassim, H., and P. Le Galès. 2010. Exploring Governance in a Multi-Level Polity: A Policy Instruments Approach. *West European Politics* 33: 1–21.

Lascoumes, P., and P. Le Gales. 2007. Introduction: Understanding Public Policy through Its Instruments? From the Nature of Instruments to the Sociology of Public Policy Instrumentation. *Governance* 20: 1–21.

Lepori, B., E. Reale, and E. Primieri. 2017. Public Funding Country Profile France Annex 11.

Lindblom, C.E. 1959. The Science of "Muddling Through". *Public Administration Review* 19: 79–88.

Linder, S.H., and B.G. Peters. 1989. Instruments of Government: Perceptions and Contexts. *Journal of Public Policy* 9: 35.

Mahoney, J., and K.A. Thelen. 2010. *Explaining Institutional Change: Ambiguity, Agency, and Power*. Cambridge: Cambridge University Press.

March, J.G., Olsen, J.P., 2005. Elaborating the "New Institutionalism". Working PaperNo.11, March 2005. Arena Centre for European Studies, University of Oslo.

Mazzucato, M. 2013. *The Entrepreneurial State: Debunking Public vs. Private Sector Myths*. London: Anthem Press.

Merton, R.K. 1938. Social Structure and Anomie. *American Sociological Review* 3: 672–682.
Ministère de l'Education nationale de l'Enseignement supérieur et de la Recherche. 2004. Rapport des Etats généraux de la recherche.
Musselin, C. 2017. *La grande course des universités*. Paris: Sciences Po Presse.
Musselin, C., and C. Paradeise. 2009. France: From Incremental Transitions to Institutional Change. In *University Governance: Western European Comparative Perspectives*, ed. Catherine Paradeise, Emanuela Reale, Ivar Bleiklie, and Ewan Ferlie, 21–49. Dordrecht: Springer Verlag.
Mustar, P., and P. Larédo. 2002. Innovation and Research Policy in France (1980–2000) or the Disappearance of the Colbertist state. *Research Policy* 31: 55–72.
Nascia, L., & M. Pianta. 2013. ERAWATCH Country Reports 2013: Italy.
Nauwelaers, C., and R. Wintjes. 2008. *Innovation Policy in Europe: Measurement and Strategy*. Cheltenham: Edward Elgar.
Nuvolari, A., and M. Vasta. 2015. The Ghost in the Attic?: The Italian National Innovation System in Historical Perspective, 1861–2011. *Enterprise and Society* 16: 270–290.
OECD. 2005. Governance of Innovation Systems. OECD.
———. 2014. Reviews of Innovation Policy: France 2014, OECD Reviews of Innovation Policy. OECD.
Onida, F., and F. Malerba. 1990. *La Ricerca Scientifica*. SIPI.
Pelkonen, A., T. Teräväinen, and S.-T. Waltari. 2008. Assessing Policy Coordination Capacity: Higher Education, Science and Technology Policies in Finland. *Science and Public Policy* 35: 241–252.
Peters, B. Guy. "Managing horizontal government." *Public administration* 76, no. (2) (1998): 295–311.
Peters, B.G. 2000. The Politics of Tool Choice. In *The Tools of Government: A Guide to the New Governance*, ed. L. Salamon, 1st ed. Oxford University Press.
Potì, B., and E. Reale. 2011. ERAWATCH Country Report 2011: Italy.
Reale, E., and B. Potì. 2009. Italy: Local Policy Legacy and Moving to an 'In Between' Configuration. In *University Governance: Western European Comparative Perspectives*, ed. Catherine Paradeise, Emanuela Reale, Ivar Bleiklie, and Ewan Ferlie, 77–102. Dordrecht: Springer Verlag.
Sabatier, Paul A. 1986. "Top-Down and Bottom-Up Approaches to Implementation Research: A Critical Analysis and Suggested Synthesis." *Journal of Public Policy* 6 (1): 21–48.
Salamon, L. 2002. *The Tools of Government: A Guide to the New Governance*. Oxford University Press.
Schneider, A., and M. Sidney. 2009. What Is Next for Policy Design and Social Construction Theory? *Policy Studies Journal* 37: 103–119.

Schoen, A., G. Carat, and N. Jan. 2008. ERAWATCH Analytical Country Report 2007: France.

Scholten, P., E. Collett, and M. Petrovic. 2017. Mainstreaming Migrant Integration? A Critical Analysis of a New Trend in Integration Governance. *International Review of Administrative Sciences* 83: 283–302.

Skodvin, T., A.T. Gullberg, and S. Aakre. 2010. Target-Group Influence and Political Feasibility: The Case of Climate Policy Design in Europe. *Journal of European Public Policy* 17: 854–873.

Spence, R.E. 2014. Italy. In *Comparative Public Administration*, ed. J.A. Chandler, 2nd ed. London: Routledge, Taylor & Francis Group.

Stoppino, M. 2001. *Potere e Teoria politica.* Giuffre.

Théry, J.-F. 2004. Les spécificites juridiques de l'administration et du pilotage de la recherche. *Rev. française d'administration publique* 112: 659.

Traù, F. 2009. Logica di un sistema di sostegno alle imprese. La politica industriale italiana dal secondo dopoguerra a "Industria 2015". *Economia e Politica Industriale.* https://doi.org/10.3280/POLI2009-004006.

Trein, P. 2017. A New Way to Compare Horizontal Connections of Policy Sectors: "Coupling" of Actors, Institutions and Policies. *Journal of Comparative Policy Analysis: Research and Practice* 19: 419–434.

Van Asselt, H., T. Rayner, and Å. Persson. 2015. Climate Policy Integration. In *Research Handbook on Climate Governance*, ed. K. Bäckstrand and E. Lövbrand, 388–399. Cheltenham: Edward Elgar.

Van der Meulen, B. 1998. Science Policies as Principal–Agent Games: Institutionalization and Path Dependency in the Relation Between Government and Science. *Research Policy* 27: 397–414.

Varone, F., and B. Aebischer. 2001. Energy Efficiency: The Challenges of Policy Design. *Energy Policy* 29: 615–629.

Vedung, E. 1998. Policy Instruments: Typologies and Theories. In *Carrots, Sticks and Sermons: Policy Instruments and Their Evaluation*, ed. E. Vedung, M.L. Bemelmans-Videc, and R.C. Rist, 21–58. New Brunswick: Transaction Publishers.

Veltz, P. 2007. *Faut-il sauver les grandes écoles? De la culture de la sélection à la culture de l'innovation*, Nouveaux Débats. Paris: Presses de Sciences Po.

Wilson, C. 2000. Policy Regimes and Policy Change. *Journal of Public Policy* 20: 247–274.

Zafonte, M., and P. Sabatier. 1998. Shared Beliefs and Imposed Interdependencies as Determinants of Ally Networks in Overlapping Subsystems. *Journal of Theoretical Politics* 10: 473–505.

Zaparucha, E. 2010. Erawatch Country Report 2010: France.

Index[1]

A
Action content, 42
Actor configurations, 7, 46, 51–52, 79–81
Actor perspective, 3, 52–53, 206, 242
Actors, 2, 27, 69, 105, 154, 205
Administrative structures, 29, 38, 45, 50, 214
Analytical techniques, 81
Applied research, 5, 5n5, 9, 116, 135n39, 232n11, 242
Authority, 13, 17, 18, 27, 29, 35, 38–42, 78, 82, 88, 108n2, 113, 116, 119–121, 123n21, 132, 135, 136, 159n5, 164, 165, 167, 172, 206, 209, 213, 216, 222, 236, 241
Automatic delivery components, 81, 215, 237, 239
Automatic incentives, 173
Automaticity, 38, 43, 45, 45n7, 46, 82, 214

B
Basic research, 5, 5n5, 9, 116, 135n39, 165, 167, 171, 232n11, 242
Bassanini reform, 153, 159, 159n5, 162, 166, 171, 179, 180, 185, 187–189
Behavior, 16–18, 33, 35, 39–43, 40n4, 47, 51, 52, 77, 81, 114n12, 121, 122, 136, 174, 241
Behavioral change, 15, 17, 18, 28, 51–53, 120, 206, 226, 236
Behavioral constraints, 41
Behavioral dimensions, 43, 53
Budgetary documents, 81
Bureaucracy, 43, 45, 46, 79, 114, 164, 167, 183, 231, 239

C
Cabinet, 11, 12, 28, 46, 48, 70, 76, 81, 82, 108, 119, 166, 170, 172, 174–176, 177n17, 178, 178n18,

[1] Note: Page numbers followed by 'n' refer to notes.

T. D'haen, *Dutch Interbellum Canons and World Literature A. Roland Holst, M. Nijhoff, J. Slauerhoff*, Canon and World Literature, https://doi.org/10.1007/978-981-99-5427-8

181, 182, 185n21, 209, 212–215, 222, 224, 234, 235, 237, 238
Cabinet periods, 83, 153, 179
Case-selection strategy, 85
Case studies, 4, 15, 19, 69, 84, 86, 154, 207, 212n2, 226, 239, 242
Civil Budget for Research and Technological Development (BCRD), 130, 218
Codebook, 82, 83, 84n11, 90
Coercion, 35, 39, 42, 112, 138
Coercive shapes, 215, 237
Collaboration, 7, 10, 14, 17, 131, 132, 161–163, 174, 178, 181, 182, 186, 209, 225, 226, 228, 235, 239, 241
Commissariat Général à l'Investissement (CGI), 107, 113, 114, 114n12, 132, 132n33, 209, 222, 222n6, 224, 225, 238
Comparative methodology, 70
Competitive clusters, 88, 106, 108, 122–123, 126, 137, 138, 209, 210, 229, 232, 241
Competitive research grants, 166
Complex governance environments, 3–16
Compliance, 39, 42, 52, 213
Confindustria, 176n15, 183, 184, 231
Consensus, 27, 185, 228
Coordinated policy strategies, 180
Coordination, 6, 7, 9, 13–16, 46–48, 50, 70, 71, 77, 78, 80, 105–107, 109, 113, 130–132, 135, 140, 153–156, 162, 171, 179, 181–183, 206, 218, 218n3, 222, 224–226, 230, 232n12, 233, 237, 238
  challenges, 3, 13, 16, 205, 225, 237
  practices, 13, 14, 77, 78, 129, 182, 218, 224, 237, 238

D
Delivery structures, 17, 48, 53, 112, 138, 163, 165, 186, 207, 215, 227, 234, 236–239, 241, 243
Design choices, 7, 8, 29, 47, 48, 76, 217
Document analysis, 82, 242

E
Economic growth, 1, 6, 8, 10, 88
Economic instruments, 39, 41, 112, 120, 123
Empirical analysis, 19, 69, 243
Encompassing R&I strategies, 108–114, 140–144, 158–166, 187–192
Executive national departments, 12, 217
Expenditure instruments, 40, 45, 110, 111, 121, 163, 164, 173, 175, 212, 213, 215, 237
Experimental development, 5, 5n5, 171, 180, 218
Expert interviews, 81–83
Exploratory framework, 47, 48, 81
External governance, 27
External stakeholders, 118, 226

F
Fiscal incentives, 164, 173, 174
Fiscal instruments, 42n5, 111, 112, 124, 174, 175, 177, 186, 212, 213, 215, 216, 237
France, 3, 4, 18, 40, 69, 72–75, 80, 81n4, 84–86, 88, 105–151, 153, 157, 169, 172, 206–217, 219–220, 224, 225, 227, 228n9, 229, 234, 235, 237–239
Fund for Applied Research (FAR), 161, 165, 174, 180

Fund for Enterprises Responsibilities, 130
Fund for Technological Innovation (FIT), 159, 161, 174, 180
Funding agency, 117, 119, 139, 166, 167, 171, 185, 230
Funding system, 105, 108, 117, 134, 135, 139, 164

G

Governance, 2–8, 12, 16, 27, 28, 34, 49–51, 72, 77, 83, 83n7, 85–90, 105, 107, 109–111, 113–115, 117, 118, 139, 155, 157–160, 163, 166–168, 170n10, 171, 182, 183, 185, 206, 209, 211, 217, 225, 230–232, 236, 239
  agreements, 8, 47, 51
  configurations, 16, 205
  systems, 6, 7, 18, 74, 117, 155, 159, 168, 170n10, 182, 224, 238
Governments, 2, 4, 8, 11, 13–18, 28, 29, 34, 35, 38–43, 46, 47, 49–53, 70–77, 79, 82, 83, 85, 87–89, 108, 112, 115–117, 118n14, 129, 135–137, 139, 154, 162, 164, 165, 168, 170, 174–176, 176n16, 178n18, 181, 182, 184, 205, 206, 209–215, 217, 226, 230, 232–234, 236, 237, 240–243
Grand societal challenges, 6–8, 10, 71
Grants, 42, 42n5, 43, 113n9, 136, 166, 169

H

Health care policies, 12, 217
High Council for Scientific Research, 130n31
Higher education, 73, 109, 113, 115–119, 130, 134n38, 153, 157, 162–163, 180, 184, 210, 211, 218, 220, 227, 230
Holistic R&I interventions, 217

I

Implementation, 2, 5n3, 7, 14, 15, 29, 43, 51, 74, 75, 81, 107, 109, 112–114, 113n10, 122, 126, 131, 132, 133n34, 135, 137–139, 149, 153, 161–166, 168–170, 174–176, 180, 192, 214–216, 222, 224, 226, 236, 237, 240, 241
Imposed autonomy, 86, 168, 184
Incubation and technology transfer services, 120
Industrial actors, 13, 14, 115, 119–129, 158, 164, 165, 172–179
Industrial development, 173, 216
Industrial districts, 157, 164, 177–179, 178n18, 184, 209
Industrial policy, 4, 108, 122, 179, 243
Industrial research, 109, 159–161, 232
Industrial sector, 88, 110, 155, 158, 178, 216, 240
Industrial strategies, 179
Information instruments, 42, 122, 163, 212, 213
Innovation, 1, 2, 5, 5n2, 5n3, 6, 9, 11, 12n7, 13, 14, 71–73, 75, 86, 87, 106, 107, 111–115, 119, 120, 122–127, 129, 132n33, 137–139, 137n42, 137n43, 153–155, 158–165, 173–176, 178, 182–184, 186, 189–191, 216, 218, 232–236, 238, 239, 242

Innovation Act, 110, 111, 228
Innovation Plan, 105, 111–112, 115, 132, 137, 138, 142–143, 162, 163, 209, 224, 228
Innovation policy, 4, 6, 87, 88
Innovation process, 5, 5n3, 9, 72, 165, 180, 242
Innovation projects, 107, 114, 124, 137n43, 163–165, 173, 176, 207
Innovation system, 4, 13, 69, 73, 119, 153, 154, 176, 240
Institute for Industrial Reconstruction (IRI), 172
Institutional coordination, 16, 50–52, 165, 182, 205
Institutional intermediation, 3, 207
Institutional landscape, 105
Institutional layouts, 4, 13, 19, 48, 69, 78, 81, 225
Institutional settings, 46
Institutional shapes, 12–14
Institutional structures, 3, 13, 16, 40n4, 70, 77, 78, 122, 205, 218, 227, 237, 238
Institutions, 3, 7, 12, 14, 15, 18, 19, 27, 29, 35, 40n4, 48, 50, 51, 70, 72, 77, 78, 81, 83, 87, 88, 107, 108, 114, 117, 119, 133, 133n34, 134n36, 137, 137n43, 153, 155–159, 161, 163, 165, 168–174, 179–184, 206, 208, 210, 217–225, 228n9, 231, 232, 232n12, 237
Instrument delivery, 17, 18, 39, 43, 45, 53, 82, 138, 206, 214, 215, 235, 237, 241, 243
Instrument families, 17, 18, 40, 41, 43, 53, 82, 206, 212, 213, 215, 237, 241, 243
Instrument implementation, 186, 226, 234, 236, 238
Instrument mix, 4, 11, 12, 14, 16, 18, 34, 35, 39, 46, 48, 69, 74–76, 78, 81, 85, 86, 107, 110, 111, 114, 117, 120–123, 134, 138–146, 150–151, 155, 158, 160, 162, 164, 165, 177, 186–201, 207, 210, 211, 215–217, 226, 227, 234, 237, 238
Instruments, 3, 4, 11, 13, 15–19, 28–35, 38–40, 40n4, 42, 42n5, 43, 45–48, 52, 53, 69–90, 109, 111, 111n8, 112, 114, 117, 120, 121, 123n21, 124–126, 129, 130, 138, 159–168, 171, 171n11, 172, 174–176, 182, 184–187, 205–207, 211–217, 225–227, 234–237, 239–243
Instrument shapes, 17, 18, 39, 42–43, 43n6, 52, 53, 122, 138, 206, 207, 213, 240, 241, 243
Instrument typology, 17, 19, 107, 155, 212, 236, 241, 242
Interest configuration strategies, 207
Interest intermediation, 3, 16, 46–48, 51–52, 69, 70, 79–81, 85, 205, 231, 233, 234, 238
Interest representation, 183–185, 226–236
Interministerial Committee for Economic Planning (CIPE), 155, 160, 160n7, 166, 180–182, 222, 224, 238
Interministerial Mission for Research and Higher Education (MIRES), 131, 134, 136, 218, 220, 221, 221n5
Internal governance, 27, 118, 157, 167–169, 211, 232
Investment for the Future Program (Plan d'Investissement d'Avenir), 106, 107, 112–114, 114n12,

120n16, 131, 132, 132n33, 135, 136, 138, 139, 143–144, 209, 222, 222n6, 225, 236
Italy, 3, 4, 18, 40, 69, 75, 76, 80, 81, 81n4, 82n5, 84, 85, 88, 153–199, 206–217, 219–220, 222, 224, 233–235, 237–239

K
Knowledge development, 1, 5, 158n2, 217
Knowledge production, 3, 5, 6, 13, 106, 206

L
Latent themes, 83
Law on Innovation and Research, 105, 110–111, 115, 140–142
Law on the Freedom and Responsibilities of Universities (LRU), 117–119, 134, 210
Les Assises de l'Innovation, 109
Linear understanding of innovation process, 180, 218
Loans, 42, 46, 107, 136–138, 137n43, 175, 235, 236, 239
Loi Allegre (Innovation and Research Law), 110–112, 133n35

M
Market-based systems, 43
Market innovation, 5, 5n3
Methodology, 19, 48, 69, 81
Methods, 5n3, 72n1, 81–86, 126, 178n18
Minister, 2, 6, 8, 12, 13, 48, 50, 79–81, 85, 107, 108, 111–113, 116, 118, 123n20, 129–133, 130n31, 132n33, 137, 156–159, 161–163, 165–168, 171, 174, 175, 179, 181–183, 209, 217, 218, 220–225, 221n5, 222n6, 229, 231, 233, 236, 238, 239, 242
Minister for Higher Education and Research, 107, 121, 133, 135, 137, 156, 158, 159, 162, 180, 182, 185, 185n21, 208, 217, 218, 220–222, 224, 225, 232, 233
Ministerial coordination, 13, 14, 77, 129, 180, 181, 219–220, 224
Ministerial organizational layouts, 14, 47, 48, 50–51, 70, 77–78, 81, 108, 180, 217–225
Ministerial R&I coordination, 129–138
Ministerial reunification, 180
Ministerial specialization strategies, 78, 158, 179, 218, 237
Minister of higher education, 129, 132, 156
Ministry for Economic Development (MISE), 85, 155, 156, 158, 160–166, 160n7, 175, 177, 179, 180, 182, 185, 223
Ministry of Economy, Finance, and Industry, 107, 109, 129, 137, 149, 229
Ministry of Higher Education and Research (MIUR), 107, 108, 119, 130, 133, 155, 156, 158–161, 160n7, 175, 178, 179, 185, 209, 210, 223
Ministry of Industry, 108, 130, 179
Ministry of Productive Activities, 179
Ministry of Public Education, 167, 179
Ministry of State-holdings, 179
Ministry of University and Research, 165–167, 169, 171

Ministry without Portfolio, 167, 179
Mission-oriented, 72–74, 76, 85, 105, 158, 165, 166, 207, 209, 215, 237
Multiannual national research plans, 155

N
National competitiveness, 6, 86
National consultation, 109, 111, 116
National Council for Research/ National Research Council (CNR), 156, 158, 158n2, 166, 171, 180, 185, 230, 233
National Fund for Science (FNS), 130, 135, 232n11
National governance, 3, 107–108, 155–159, 167, 179, 210, 222, 224, 228, 238
National Hydrocarbons Board (ENI), 172
National innovation systems, 70, 71, 157
National Institute of Nuclear Physics (INFN), 156, 158
National reform of public administration, 159
National Research Program (NRP), 157, 160, 180, 181, 222
National Strategy for Research and Innovation (SNRI), 107, 112, 131, 222, 228

O
Opportunity structures, 15, 46, 47
Ordinary Financing Fund (FFO), 157, 168, 170, 184
Organizational layouts, 7, 47, 69, 81n4

P
Partisan politics, 4, 11, 48, 81
Partisan politics orientations, 16, 48, 81, 205, 236
Path dependency, 76, 139, 215–217
Performance evaluation, 168
Policy capacity, 182
Policy changes, 4, 8, 11, 18, 48, 52, 86, 121, 140, 167, 177–179, 227, 230, 243
Policy choices, 28, 88, 89, 140, 215, 222, 237, 239
Policy design, 1–19, 28–53, 69, 70, 77, 78, 82, 84, 86, 88, 89, 105–151, 153–199, 205–244
  choices, 3, 8, 14–18, 27–53, 70–81, 181, 205, 211, 212, 238
  theories, 7, 40
Policy implication, 19, 207, 242–244
Policy instruments, 3, 8, 11, 17–19, 28, 29, 34–49, 43n6, 51–53, 53n8, 69–72, 76, 77, 79, 82, 89, 114, 129, 140, 178, 186, 206, 211, 212, 214–216, 234–238, 240–243
Policy instrument selection, 29, 48, 236, 241, 244
Policy issues, 19, 27, 47, 129, 133, 184, 236, 239
Policy legacy, 89, 119
Policymakers, 1–18, 27–29, 34, 35, 38, 41, 42, 47–49, 51–53, 69–71, 76, 77, 79, 80, 83, 88, 89, 109, 112, 114, 115, 118–123, 133n35, 137–139, 153, 160, 162–165, 167, 168, 170, 173, 177, 180, 184, 205–208, 210, 211, 216–218, 226, 228–231, 233–236, 238, 239, 241–243
Policymaking, 3–16, 19, 27, 29, 35, 48, 50, 51, 70, 78, 79, 81, 83, 84, 86, 88, 107, 109, 129–132,

130n31, 140, 158, 163, 165, 179–184, 186, 205, 206, 217, 218, 222, 224–226, 229–231, 236–238, 242, 243
Policy mix, 4, 13, 18, 19, 34–47, 52, 86, 105, 106, 121, 122, 124, 127, 129, 138, 146–150, 163, 171–173, 182, 186, 206, 207, 216, 225, 236, 241
Policy preferences, 8, 15, 226
Policy recipients, 3, 10, 13–16, 18, 27, 40, 41, 47–49, 51, 70–71, 76, 79, 81n4, 161, 206, 216, 226, 242
Policy responsibilities, 7, 12, 14, 217, 218, 222, 225, 237
Policy responsiveness, 3
Policy sectors, 2, 4, 6, 7, 10, 12–14, 29, 47, 167, 208, 226, 244
Policy taker, 3, 4, 8, 14–19, 29, 47, 48, 51–53, 83, 109, 110, 205, 206, 210, 215, 216, 226, 227, 233, 238
Policy types, 48–49, 69, 71–77, 84, 85, 215, 216, 237
Political coalitions, 212, 215, 237
Political institutions, 27, 29, 52
Political orientations, 12, 19, 47–49, 81, 131, 178, 205, 211–215, 236, 237
Political payoffs, 70
Political polarization, 214
Political priorities, 3, 11–13, 28, 46, 50, 70, 78, 182, 211, 216, 217
Political process, 11, 34, 69, 155, 207, 241
Political relations, 43, 226
Political turnovers, 215–217
Politics of instrument choices, 206, 211–217
Politics of policy design, 8–12
Private investments in R&D, 119
Private R&I performers, 108, 127, 136–138, 228, 234
Process innovation, 5, 5n3, 9, 72, 165, 180, 242
Product innovation, 5, 5n3
Public activities, 5, 6
Public actors, 9, 14, 51, 72, 79, 111, 173, 235
Public agencies, 2, 6, 12, 13, 40n4, 43, 45, 46, 79–81, 85, 106, 112, 114, 120, 132–136, 140, 160, 169, 231, 232, 234, 238
Public funding, 16, 120, 134, 139, 167, 185, 211
Public investment, 2, 72, 75, 113
Public research, 2, 6, 9, 15, 80, 85, 105, 108–111, 110n7, 114–119, 121, 130–131, 133–136, 134n38, 139, 153, 155, 156, 158–161, 163, 166–175, 178, 179, 184, 185, 193–194, 210, 227–233, 232n12, 238
Public research organizations (PROs), 5, 46, 72, 83n8, 105, 107–109, 115–117, 119, 120, 123, 125n24, 126, 129, 130, 133–136, 134n38, 135n39, 139, 154–157, 158n2, 164–167, 171–172, 186, 207, 210, 218, 220, 221, 223, 227, 229, 230, 232, 232n11, 232n12, 236, 241
Public support for R&D, 119

Q

Qualitative Content Analysis (QCA), 82

R

R&D funding, 117, 154
R&I, *see* Research and Innovation

R&I instruments, 10–12, 14, 16, 19, 49, 69, 70, 83–86, 107, 123, 155, 158, 160, 180, 186, 206, 211–214, 216, 237, 239
R&I mix, 226, 238
R&I performers, 4, 15, 18, 46, 75, 80, 81, 106–108, 112, 138, 139, 158, 163, 182, 186, 207, 225–230, 233–236, 238, 239
R&I policies, 2–19, 27, 29, 46–53, 69–90, 105–151, 153–199, 205–244
R&I policy-type traditions, 215
Rapport Guillaume, 105, 109
Recruitment system, 167
Regulation, 40, 40n3, 87, 88, 110n7, 138, 163, 167–169, 171, 172, 178n18, 184, 211–213
Regulation instruments, 212, 213
Research Act, 106, 112, 115–119, 145–146, 209, 227, 230, 231
Research and Innovation (R&I), 1–19, 27, 29, 43, 43n6, 46–53, 69–90, 105–151, 153–199, 205–244
Research and Innovation strategy, 2, 3, 8, 16, 72, 81n4, 84, 87, 88, 106, 110, 112, 132, 163, 175, 181, 186, 205, 223, 225, 227, 230, 235, 237, 239
Research & Innovation systems, 1, 72, 108, 109, 113, 122, 125, 161, 162, 172, 176, 185, 208, 209, 228–230
Research companies, 158
Research design, 81–86
Research funder, 166, 230, 233
Research policies, 4, 6, 34, 87, 130n31, 156, 162, 218, 230
Research projects, 78, 135, 158, 158n2, 159, 161, 166, 167, 178, 233
Research Tax Credit (CIR), 110, 124–127, 129

S
Science policy, 4, 15
Sectorial missions, 166, 173
Sectorial public research organizations, 130
Semi-structured interviews, 82, 83
Small and medium enterprises (SME), 75, 87, 106, 107, 109, 111–113, 119–124, 123n21, 126, 127, 135n39, 136–139, 137n42, 153, 154, 157, 160, 162–164, 173, 174, 176–178, 183, 209, 216, 228, 229, 231, 232, 232n11, 232n13, 235, 241
Special Applied Research Fund (FISR), 158, 159, 167
Stakeholders, 1, 6, 7, 15, 18, 29, 34, 43, 45, 47, 53, 70, 71, 81, 83, 109–111, 115, 120–123, 130n31, 131, 133, 135, 138, 172–176, 180, 185, 214, 226, 227, 229n10, 230, 232, 232n12, 233, 235, 236, 238, 242
State holding enterprises, 172
State participation, 172, 207
State-run-enterprises, 173
Strategic behaviors, 206, 226
Strategic research projects, 166

T
Targeted funding, 123, 173
Target groups, 16, 42, 52, 79, 80, 117, 206, 231, 241
Target population, 4, 8, 15, 17, 18, 27–29, 34, 39, 41–43, 46, 48, 51, 53, 79–81, 180, 206, 207, 212, 213, 226, 231, 234, 238–240, 242, 243

Taxation instruments, 41
Tax credit, 125–127, 129, 173, 229, 241
Tax credit for research, 124–129, 220, 241
Tax incentive, 124, 171–173, 220
Technological districts, 164, 178, 179, 186, 209
Technological innovation, 10, 13, 14, 87, 111, 158–160, 163, 173
Technological networks, 120
Technological policy, 4
Technological Research Fund (FRT), 130, 135, 232n11
Technology transfer, 74, 85, 105, 109, 121, 162, 163
Territorial and industrial development, 176
Thematic Analysis methodologies, 82–83
Thematic research, 120, 166

U

Universities, 5, 10, 15, 46, 74, 105, 108–110, 108n2, 115–119, 123, 125n24, 126, 129, 133, 133n34, 134, 134n36, 135n39, 139, 157, 158n2, 163–165, 167–170, 170n10, 172, 176n16, 179, 184, 185, 210, 211, 218, 227, 228, 228n9, 232n11
University performance, 108

V

Venture capital, 109, 110, 176

FSC
www.fsc.org
MIX
Papier | Fördert gute Waldnutzung
FSC® C083411

Kolibri 360 GmbH
Ferdinand-Jühlke-Straße 7
99095 Erfurt, Deutschland
produktsicherheit@kolibri360.de